To the

THE PIONEERS OF BRITISH COLUMBIA

1 *The Reminiscences of Doctor John Sebastian Helmcken*, edited by Dorothy Blakey Smith

2 *A Pioneer Gentlewoman in British Columbia: The Recollections of Susan Allison*, edited by Margaret A. Ormsby

3 *God's Galloping Girl: The Peace River Diaries of Monica Storrs, 1929-1931*, edited by W.L. Morton

4 *Overland from Canada to British Columbia*, edited by Joanne Leduc

5 *Letters from Windermere, 1912-1914*, edited by R. Cole Harris and Elizabeth Phillips

6 *The Journals of George M. Dawson: British Columbia, 1875-1878*, edited by Douglas Cole and Bradley Lockner

7 *They Call Me Father: Memoirs of Father Nicolas Coccola*, edited by Margaret Whitehead

8 *Robert Brown and the Vancouver Island Exploring Expedition*, edited by John Hayman

9 *Alex Lord's British Columbia: Recollections of a Rural School Inspector, 1915-36*, edited by John Calam

10 *To the Charlottes: George Dawson's 1878 Survey of the Queen Charlotte Islands*, edited by Douglas Cole and Bradley Lockner

Edited by Douglas Cole and Bradley Lockner

To the Charlottes
George Dawson's 1878 Survey of the Queen Charlotte Islands

UBC Press / Vancouver

© UBC Press 1993
All rights reserved
Printed in Canada on acid-free paper ∞

ISBN 0-7748-0415-7

ISSN 0847-0537 (The Pioneers of British Columbia Series)

Canadian Cataloguing in Publication Data

Dawson, George M., 1849-1901.
To the Charlottes

(The Pioneers of British Columbia, ISSN 0847-0537)
Includes bibliographical references and index.
ISBN 0-7748-0415-7

1. Dawson, George M., 1849-1901 – Diaries.
2. Queen Charlotte Islands (B.C.) – Description and travel.
3. Geology – British Columbia – Queen Charlotte Islands.
4. Haida Indians. 5. Indians of North America – British Columbia
– Queen Charlotte Islands. 6. Geologists – Canada – Diaries.
I. Cole, Douglas, 1938- II. Lockner, Bradley John, 1950-
III. Title. IV. Series.

FC3845.Q3D39 1993   917.11'12043   C93-091095-8
F1089.Q3D39 1993

Publication of this book was made possible by ongoing support from the Canada Council, the province of British Columbia Cultural Services Branch, and the Department of Communications of the
Government of Canada.

UBC Press
University of British Columbia
6344 Memorial Rd
Vancouver, BC v6T 1z2
(604) 822-3259
Fax: (604) 822-6083

Contents

Illustrations / vii

Acknowledgments / ix

Introduction / 3

George Dawson's 1878 Journal / 9

On the Haida Indians of the Queen Charlotte Islands / 97

Notes / 167

Index / 205

Illustrations

Following page 54

1 George M. Dawson, about 1885. BC Provincial Archives
2 Dawson among other officers of the North American Boundary Survey, 1873. National Archives of Canada (NAC) PA74675
3 'Looking east down Houston Stewart Channel from Ellen Island,' 12 June 1878. NAC PA152383
4 'Dolomite Narrows, Burnaby Strait,' 26 June 1878. NAC PA51091
5 'Volcanic agglomerates at Point 4, North Shore of Ramsay Island,' 1 July 1878. NAC PA152363
6 'Grassy patch beneath which hotspring rises, Hotspring Island,' 4 July 1878. NAC PA51103
7 'Clue's village of Tanu, Laskeek Bay,' 9 July 1878. NAC PA37753
8 'Cumshewa Indian village,' 16 July 1878. NAC PA37752, NAC PA378151
9 'Skedan Indian village,' 18 July 1878. NAC PA44329, NAC PA38148
10 'Skidegate Indian village,' 26 July 1878. NAC PA37755, NAC PA37756
11 'Basaltic columns, Tow Hill,' 9 August 1878. NAC PA1371

Following page 118

12 'Masset Indian village,' 10 August 1878. NAC PA38149
13 'Mission Buildings at Masset,' 18 August 1878. NAC PA44332
14 'Looking up Masset Inlet from Indian village,' 18 August 1878. NAC PA44327
15 'Kung Indian Village, Virago Sound,' 21 August 1878. NAC PA37757
16 'Group of Haida Indians, including Chief Edenshaw, Second from Left, Standing,' 23 August 1878. NAC PA38154
17 'Edenshaw and Hoo-ya, Chiefs of Yats at Masset,' 23 August 1878. NAC PA38147

18 'Pillar Rock, Pillar Bay, Graham Island,' 24 August 1878. NAC PA51118
19 'Dadans Indian village, North Island,' 24 August 1878. NAC PA44326
20 'Port Simpson village and church,' 30 August 1878. NAC PA44333
21 'Metlakatla church and mission buildings,' 2 September 1878. NAC PA44335
22 'Looking north up Klemtu Passage,' 9 September 1878. NAC PA152368
23 'Glaciated rocks and Indian graves, Shell Island, Fort Rupert,' 2 October 1878. NAC PA51117
24 'Indian village at Hudson's Bay Company post, Fort Rupert,' 2 October 1878. NAC PA44336
25 'Comox Wharf,' 12 October 1878. NAC PA51108

Acknowledgments

In preparing Dawson's 1878 journal for publication the editors have relied upon the expertise of a large number of people. Special thanks must be given to W.H. Mathews, whose knowledge of British Columbia's geology contributed enormously to our understanding of Dawson's geological comments. We are also indebted to a number of people for their advice on various subjects. These include Nancy J. Turner and Robert Ogilvie for botany and ethnobotany; Don E. McAllister for fish; Ralph Maud and Randy Bouchard for folklore; Susan Pyenson for L.W. Dawson; Ingrid Birker for palaeontology; Arnoud H. Stryd, John G. Foster, and Grant Keddie for archaeology; Dawson's grandniece, Mrs. Anne V. Byers, R. Garth Walker, David Mattison, and Dr. Sunil Mehra for biography. Al and Irene Whitney, aboard the *Darwin Sound II*, were graciously helpful with the marine topography of Moresby Island. We owe a deep obligation to the staff of the McGill University Archives (especially Rob Michel) and the Rare Books and Special Collections Department of the McLennan Library, McGill University. Specialized information was provided by the Provincial Archives of British Columbia (especially the Map Division) and the National Photography Collection. Other institutions which assisted were the Missouri Botanical Garden, the Royal School of Mines, the Canadian Museum of Civilization Library, the McCord Museum (especially Conrad Graham), the College Archives of the Imperial College of Science and Technology, London, and the New Westminster Public Library. Richard Mackie provided corrections and additions to the historical annotations for this edition. A number of Simon Fraser University students assisted in the project, including Don Dale, Shannon Steele, Karen Sanger, Beth Rheumer, Michelle Le Grange, Stephen Gray, Diane Dupuis, Todd Owen, Sheliza Lalji, Fred Fuchs, Robin Anderson, and Rod Fowler. Barbara Barnett Lange, while with the Dean of Arts office at Simon Fraser University, was responsible for the original textual keyboarding, and our indebtedness to her and to the dean are enormous. Margaret Sharon of Simon Fraser's Computing Services provided continued and varied assistance that eased the process of transforming the manuscript into an edited text.

Financial assistance came through Simon Fraser University's Programs of Distinction, with additional university assistance from the President's Club, the Dean of Arts, the Publications Committee, and the Historical Records Institute. The Leon and Thea Koerner Foundation and the John S. Ewart Foundation also provided assistance.

Almost all of the manuscripts used here are the property of McGill University which, by generously allowing us to publish them, made this endeavour possible.

Editorial preparation was carried out as part of the program of the Historical Records Institute of Simon Fraser University.

To the Charlottes

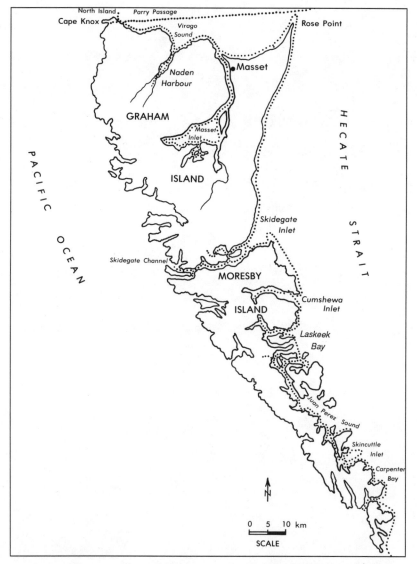

George Dawson's 1878 journey to the Queen Charlotte Islands

Introduction

Haida-Gwaii, the Queen Charlotte Islands, occupy a special place in the natural and human history of the northwest coast of North America. Detached from the mainland by eighty kilometres of water and from Alaska's southernmost Prince of Wales Island by fifty kilometres, they retain a separateness that reflects millennia of isolation. Geologists believe that they remained largely free of the Great Cordilleran ice sheet of the Pleistocene era 14,500 years ago. Although subject to their own mountain glaciation during that period, they probably provided ice-free areas that served as refuges for plant and even animal forms.

This exceptional history is reflected, for example, in the special character of the Charlottes' flora, especially their mosses. The islands harbour a number of rare or unique species, although faunal absences are more striking than floral presences. No deer, beaver, raccoon, mink, or snakes are indigenous to the islands, and they entirely lack alpine or subalpine mammals and birds.[1]

These singular isles became, at least 8,000 years ago, home to humans – predecessors of the Haida people. Arriving by sea, they lived relatively isolated from other humans until about 5,000 years ago, when microblade tool technology was introduced from the north. By 2,000 years ago, human culture on the Charlottes began to resemble more closely that of the mainland and islands to the east and north. From that time on, archaeological remains mark the Charlottes' inhabitants as identifiable forerunners of the Haida met in the late eighteenth century by European explorers and merchants. By then their culture was similar in many ways to that of their Tsimshian and Tlingit neighbours, but each group remained at least as distinct, one could probably say, as were the European nationalities of the time. In language, however, the parallel fails: the Haida language, at least, has no known relationship – aside from some words obviously borrowed from the Tlingit and the

Tsimshian – with any other language. Moreover, even noting a dialect difference between those who became known as Skidegate and Masset speakers, the Haida language is more homogeneous than any other Northwest Coast language.

Haida-Gwaii and the Haida first came to the attention of the Europeans with Juan Pérez's landfall (though neither he nor any of his crew set foot on shore) at Dadens village on North Island in 1774. The English fur trader George Dixon, who left his name to the waterway between Graham and Prince of Wales islands, gave the island group their permanent English name in 1787, after his own ship.

The Haida at Dadens and then elsewhere took to the fur trade quickly, supplying sea otter pelts in exchange for clothing, blankets, chests, copper, and iron. Early in the nineteenth century, however, the sea otter declined and the Haida and their islands were left largely alone, though the people continued their own pre-European trade across Dixon Entrance with their Kaigani Haida kinsmen and across Hecate Strait with the Tsimshian, especially for oil from the eulachon, a species of fish absent from Haida-Gwaii. They also sold potatoes to European posts on the mainland.

A short-lived gold rush on the west coast of Graham Island scarcely touched the Haida; the same was not true of the 1862 smallpox epidemic that spread northward to devastate every coastal village. The consequent depopulation contributed to major changes in Haida society. The remaining Haida began to consolidate, abandoning more remote villages in a movement that ended with only two permanent Haida centres, Skidegate and Masset. That trend was under way but incomplete when, in the summer of 1878, George Dawson, a twenty-eight-year-old Canadian geologist, undertook the first scientific survey of the islands and their people.

The Charlottes, when Dawson surveyed their eastern and northern shores that summer, were relatively untouched by European exploitation. The sea otter was nearly extinct, but other fauna and the entire islands' flora remained virtually intact except as needed by the Haida for their life and industry. The short, unproductive rush to Gold Harbour in 1851-2, abortive attempts at copper mining on Burnaby and Skincuttle islands in the early 1860s, and an equally fruitless coal venture in Skidegate Inlet a few years later were the extent of mineral exploitation. The forest was unused except for the Haida's own demands. Only a handful of Whites lived permanently on the islands. Anglican missionary W.H. Collison, with Mrs. Collison, and Hudson's Bay Company trader Martin Offut resided at Masset; entrepreneurs J. McB. Smith and Andy McGregor lived at Skidegate, where they operated a small plant, pressing oil from dogfish livers for the southern logging industry. Even these people were recent arrivals: Offut had come in 1870, the rest only in 1876. Indeed, the whole of coastal British Columbia north of Comox remained largely in Native occupancy. A few Whites – traders, missionaries, early cannerymen – lived at Alert Bay, Fort Rupert, Bella Bella, Inverness, Port Essington, Metlakatla, and Port Simpson.

The contours of the coast's bays, sounds, and river estuaries were well delineated by 1878. George Vancouver's charts of the 1790s had meticulously detailed the shores of the mainland and Vancouver Island, and eighty years of coastal navigation, including Royal Navy hydrography, had left the mainland well charted. The Charlottes, on the other hand, were much less known. Vancouver had only sighted them from a distance and Admiralty charts of the 1870s left them in broad outline, with most of their bays, channels, and small islands uncharted and unnamed. Similarly, the Haida were known more from their casual visits to Victoria and other southern settlements than from being observed in their homelands. For example, James G. Swan of Port Townsend in Washington Territory had published in 1874 a short Smithsonian Institution memoir based on discussions with Chief Kitkane of Klue and three other individuals.[2]

Such, briefly, was the state of knowledge when George Mercer Dawson sailed northward on *Wanderer* in the spring of 1878. His aim was general exploration of the Queen Charlotte Islands, especially with respect to their value for resource exploitation. He travelled as a member of the Geological Survey of Canada, and, while primarily a geologist, he was attentive to topographical and hydrographic mapping and to agricultural potential, and was also concerned with the biological nature of the islands and the ethnological character of their Haida inhabitants.

Dawson had spent the first six years of his life in Nova Scotia, where he had been born in Pictou on 1 August 1849. His Nova Scotia-born father, William (later Sir William) Dawson, had studied at Edinburgh University, where he had met his future wife, Margaret Mercer, and then had begun a career as an educator, first at Dalhousie, then as Nova Scotia's superintendent of education, and finally, in 1855, as principal of Montreal's McGill College. All the while, however, the senior Dawson continued his youthful interest in geology, an avocational career that was forwarded by the visits of the eminent British geologist Sir Charles Lyell and by his own publication of *Acadian Geology* in the same year that he assumed the principalship at McGill College.

Residing on the spacious McGill campus from the age of six, the young George Dawson enjoyed a rich childhood. He tended his garden, collected butterflies and plants on the mountain behind his college home, helped his father in the college museum, and indulged in the usual pastimes of boys his age. Then, by the winter of 1858-9, while in his ninth year, he began to experience the effects of a spinal illness that would permanently affect his life.

He was suffering from Pott's disease, or tuberculosis of the spine, a slow-working and painful disease that caused the affected vertebrae to soften and collapse and the spine to twist and curve. He was treated with a body truss in Boston and survived the threat of paraplegia and a roughly 30 per cent chance of death, but he remained for several years a bedridden invalid suffering from headaches and pain in his back and limbs. What was just as serious was that his growth was stunted and his upper body permanently deformed. He never attained more than the stature of a ten-year-old and always had the bulky torso of a hunchback.

George received lessons at home from his father and tutors as his health gradually improved, and, as he gained strength and stamina, his headaches decreased. He continued to develop his naturalist interests and, by 1865, was able to accompany his father to Britain. At nineteen, he enrolled in classes at McGill, then decided to follow in his father's geological footsteps by taking the three-year program in geology and mining offered by London's Royal School of Mines. The strength of the British school, in which Dawson enrolled in 1869, was its faculty. Many of the pre-eminent names of nineteenth-century British geology, mining, chemistry, and palaeontology taught there, including T.H. Huxley.

After an indifferent first year, Dawson excelled, earning the school's highest awards. Upon graduation in 1872, he decided to return to Canada and to dedicate himself to geology, not to mining. He had hoped to obtain a position with the Geological Survey of Canada, but his father insisted that he accept an opening as naturalist and geologist with the joint British and American survey of the international boundary from Lake Superior to the Rocky Mountains. Fortunately, the Geological Survey kept a position open until Dawson finished his boundary work two years later. Thus, on 1 July 1875, he was appointed a geologist with the Montreal-based Survey. He arrived in British Columbia a month later. His first two summers were devoted to exploration of the central interior west of the Fraser River in association with the government's Canadian Pacific Railway surveys. In 1877, he turned his attention to the Thompson-Shuswap region and the Fraser, Similkameen, Okanagan, and Nicola valleys of the southern interior. In 1878, he shifted to the coast, examining northern Vancouver Island and the Queen Charlotte Islands.

Dawson's exploration brought to the Charlottes one of the most remarkable field scientists of the nineteenth century. His geological work on the western prairies, in the British Columbia and Yukon mountains, and along the Pacific coast laid a firm basis for the glacial history of the Cordilleran region and for an understanding of its stratigraphic complexity. Along with this ability to synthesize from field experience, Dawson possessed the finest attributes of a practical geologist – the ability to inventory coal and mineral deposits and prospects and to analyze current mining activities. As well, he drew maps, collected fossil, plant, and insect specimens, and, most significantly, investigated the ethnology of British Columbia's Native peoples, with published papers on British Columbia languages, on the Interior Salish, the Kwakiutl, and, perhaps his best work, on the Haida.

Dawson's personality and temperament barely emerge from his letters and journals, but he was in many ways happiest when in the field. In London he had accepted invitations grudgingly. 'I wish you would not stir up these people to ask me out,' he had written his mother, 'as I hate going out, and dinner especially.' In Victoria, he socialized with the city's first families, but insisted that 'there are *very* few people here whom I care to know well.' In Montreal in 1880, the first summer in five years not spent in the field, he found the city less congenial than the bush. 'I cant say I find being here in the hot weather at all equal to being in the woods, not that I feel the heat, but the solitude is more oppressive because less natural.'[3]

His physical appearance undoubtedly affected Dawson's social relations. 'Keenly sensitive as he was by nature,' noted an acquaintance, 'he felt his deformity deeply and abstained from going into that society which his mental gifts and graces fitted him to adorn.'[4] In none of his letters or journals is there a single allusion to his physical condition. His deep reserve, on the other hand, could merely have been characteristic of the male side of the Dawson family. Even George found Rankine, his youngest brother, a quiet companion on the *Wanderer* voyage. Among acquaintances and friends, however, Dawson could assume a cheerful geniality, expressed in witty sallies, good-natured badinage, and a quick, wise smile. But, although respected and liked by most of his colleagues, he made few, perhaps no, intimate friends. He lived at the family home in Montreal and, after the Survey moved to Ottawa in 1881, in rented rooms. He never married.

Dawson continued his fieldwork year after year, including a memorable journey in 1887 through the Yukon Territory (Dawson City bears his name), though promotion to assistant director, then in 1895 to the directorship of the Geological Survey, increasingly impinged upon his field seasons. He died unexpectedly on 2 March 1901. Acute bronchitis, against which his restricted lung capacity gave him inadequate resistance, was the cause.

This edition of Dawson's 1878 exploration of the Queen Charlotte

Islands is drawn from *The Journals of George M. Dawson: British Columbia, 1875-1878*[5], with the addition of his article 'On the Haida,' published originally as an appendix to the Geological Survey of Canada's *Report of Progress for 1878-79*.[6]

The journals and family letters which form the first portion of the volume are from the manuscripts now held by the McGill University Archives (MUA). They are reproduced as faithfully as print allows. Thus capricious capitalization, odd misspellings, and occasionally incomplete punctuation are included. Dawson usually kept his daily field journal entries on the right-hand page, reserving the left-hand page for thematic comments on subjects such as ethnology or glaciation. This left-hand-page material is indicated by [verso] and [end verso] and placed as appropriately as possible.

The following editorial symbols have been used:

Blank space in manuscript	[blank]
Material cancelled in manuscript	<....>
Material inserted in manuscript	{....}
Illegible reading	[....]
Conjectural reading	[conjecture?]
Editorial comment	[comment]

'On the Haidas' is reproduced without annotation. Its value resides in its description of Haida customs as Dawson learned about them during his trip, partly from a few Haida and partly from the missionary Collison and others. His understanding is necessarily imperfect, especially of religious beliefs and ceremonial practices. The account is tinged with the almost inevitable condescension of a white, nineteenth-century observer. Readers must view the essay with caution, as a document that tells us a great deal about the Haida but also something about Dawson's own assumptions and prejudices.

George Dawson's 1878 Journal

[Dawson left Montreal on 30 April by rail via Toronto and Port Huron to Chicago. He travelled with his youngest brother, Rankine. The reason for Rankine's presence is unclear, though the Dawson family probably hoped that a summer season with George would benefit the youth, whose unsteady character was already a problem. They arrived in San Francisco on 8 May and at noon on 10 May left for Victoria on the steamer *Dakota*. Dawson used a small pocket journal until his departure from Victoria on 27 May 1878, when he began his main 1878 journal.]

May 13. Entered Esquimalt harbour at 6.30 A.m. after a remarkably quick & pleasant passage. Got Carriage over to Victoria & put up at Driard House. Find Torrence[1] still here. Get luggage up Go about various matters of business. Enquire about Craft. Find Douglas[2] not yet here but had given him time to the 15th. Enquire about other Craft in case he does not turn up.

G.M. DAWSON TO MARGARET DAWSON,[3]
13 MAY 1878, VICTORIA

Here we are in Victoria at last, after a remarkably fast & pleasant passage from San Francisco, & an uneventful railway journey over the west. The steamer brought us in here early this morning, So early that we had to content ourselves with a cup of coffee on board & come on to Victoria for breakfast. Fraser Torrance is still here, leaves for the interior next friday to visit first – I believe – Cherry Creek.

Today I have not been able to more than go round making enquiries. Capt. Douglas with his schooner is not yet here. He has still a day to the date at which I appointed to meet him, however, & if he does not turn up in time I think I shall have little difficulty in procuring a suitable Craft here from someone else. Meanwhile it will take a day or two to get other preparations made & plans more definitely laid.

Capt. Lewis[4] of the HB Co who has Sailed much on the Northern Coast Says the natives of the Queen Charlotte Islands are not to be feared & also advises me to go up there at once as the weather in the spring & early Summer is best. He says that when once the wet rough weather begins in autumn it generally lasts on through the winter. Rankine[5] begins already to look sunburnt & well though still thin enough. He is a singularly reticent travelling Companion & when anything particularly pleases him only Chuckles internally or whistles. I hope to improve him during the summer.

When in San Francisco we went to the Chinese Theatre, where I had not been before. It was well worth seeing for once, though the music, composed of peculiarly reedy & squaky notes, mingled with tones of horns & gongs, was anything but Enjoyable. The first part of the performance was a comedy, which to those who did not understand what was being Said, seemed long drawn & stupid, especially as much of the conversation was Carried on in a sort of sing-song. After this Came to a happy end an exhibition of tumbling began, which was quite wonderful. The dresses throughout were wonderfully gay with silver gold & silk. The audience was almost altogether of Chinamen, who seemed to enjoy the display heartily. The atmosphere was rather dense with tobacco smoke before we left, which was at about 11 P.m. & how much longer the fun lasted I do not Know. In a day or two I will be able to write giving Plans more in detail, till which time excuse this little scrawl & Believe me.

G.M. DAWSON TO BERNARD JAMES HARRINGTON,[6]
13 MAY 1878, VICTORIA

The quartz excitement[7] still Continues here, though the present aspect of affairs is that all the available Capital of the place having been absorbed in various more or less precarious speculations. Stock jobbing is at a standstill. From various sources I hear of assays Made in San Francisco & elsewhere, for private parties, the result of which was more favourable than Any tried at the survey. If you can find time to make trial of the samples lately forwarded, before long, I feel Sure it would be appreciated here by the Govt. people.

The proof of my map,[8] corrected as well as I can do it, was sent back from Toronto. I hope you have received it safely. There were errors both in the black & in the colours.

In the copy of the report I brought with me, the memorandum, –

which was to be printed on a separate slip and inserted – relating to the two maps of which the publication is postponed, were omitted. This was probably accidental.

In leaving I forgot to say that I should like to have the box of rock specimens which Mr. Weston[9] prepared shipped as ordinary freight during the summer. July would be soon enough, as I may not find time to do anything with it till my return here in the Autumn. The Casella thermometer[10] if another comes out, should be forwarded, by post if possible, to save expense, to my address Care C.P. Ry. office Victoria. I will try to Keep the people there posted as to our whereabouts.

Enclosed is a slip with a few addresses to which Mr. White[11] might forward Copies of the Rep. of Progress[12] when published in complete form.

Please excuse me for troubling you with these little things. Love to Anna,[13] to whom I must Soon write – & to Eric.[14]

May 12-17. Victoria looking after equipment & making preparations. Douglas not arriving on 15th make arrangements on following day with Capt. Sabotson[15] for schooner Wanderer. Evg. of May 17. dine at Mr. McDonalds.[16]

Sunday May 19. Writing & Pm for a walk by Beacon Hill & Ogden Pt.

G.M. DAWSON TO ANNA LOIS HARRINGTON, 19 MAY 1878, VICTORIA

I still date, worse luck, from this place, but hope to get away this week, though the precise day is yet uncertain. When we got here, after a remarkably quick & pleasant passage from San Francisco, Capt Douglas & his Craft were not to be heard of, but as I had given him till the 15th I was obliged to wait till that date, making enquiries in the meantime as to other available Chances. On the 16th I made an arrangement for the schooner *Wanderer* of about 20 tons & as soon as she is ready for sea will be off. She belongs to a man Called Sabotson who is getting new sails for her <....> & as Soon as these are ready all will be complete. Since engaging the Wanderer I have heard that Douglas' schooner which Mr. Richardson[17] formerly had was blown ashore in a squall somewhere near Comox & so much damaged that she Cannot be put in repair for some time. The *Wanderer* is I think a good Craft, with plenty beam, & built originally for a Pilot boat. If the Captain & Crew, the latter Yet to be selected, turn out well all will be right.

There are as you may suppose innumerable little things to look after here & it would seem that we Cannot even get a clear half day for any work about this place. Tomorrow I shall make a vigorous effort to clear off the remainder of my list, so that any days remaining may be free from little engagements Rankine Comments on the fewness of good looking girls in this part of the world, especially in contrast with their abundance in San Francisco. The best looking & therefore most highly esteemed he has yet encountered is a Miss McDonald,[18] daughter of the senator of the same name who travelled over the ry. with us & asked us to his home the other night to dinner.

 The weather is fine & settled looking now, though there have been a few windy Cold days since we landed. I long to be off taking advantage of the fine time. When all is ready, following advice, I shall strike straight away for the Queen Charlottes & after spending as much of the summer there as the Country seems to warrent & weather permits, shall begin to work back toward Vancouver. I enclose a little memo. for Mamma.[19]

May 20. Business of various Kinds in connection with outfit
May 21. Morning about town on business. P.m. walk to Limestone exposure at Indian reserve Esquimalt & hunt for fossils. Evg. walk out to McDonalds.
May 22. Morning as usual going about numerous little affairs. P.m. go by invitation to Croquet at Gov. Richards[20] stay for dinner at which Capt & Mrs Robinson[21] & Miss Dupont.[22]
May 23. Enquired about Sails which it now appears Cannot be finished before Saturday. Had hoped to go off on Friday. Visit schooner which now nearly ready. Tell Mr Charles[23] he may have the use of the boat on the Queens Birthday.

 P.m. for a drive & scramble to top of Cedar Hill[24] with Mr McDonald. Evg. row on the harbour.
May 24. Queens birthday holiday making everywhere interfering with work on Sails &c. Morning walked to Beacon Hill & Clover Pt to see Base Ball match & rifle firing. P.m. up 'the arm' with a party of Mrs Robertson[25] including Duponts, Powells, Poolies[26] &c. to see the races & regatta. The Indian Canoe races specially interesting from the excitement & vim with which the competitors went to work. Got back about dark. Evg. Called on Mrs. Bowman.[27]
May 25. Find it will be impossible to get away on Sunday morning as had hoped owing to non completion of sails & Carpenter's work. Make arrangements which I think will secure completion of work & enable us to get away on Monday morning. Get supplies Shipped down to wharf & preparations as far as possible completed. Evg. dine with Keith[28] & a party of friends.[29]

G.M. DAWSON TO BERNARD JAMES HARRINGTON,
26 MAY 1878, VICTORIA

Your note of May 7th enclosing copy of petition as to Surveys of Mineral lands etc. of B. Columbia, arrived yesterday[30] If the govt. think of doing anything in the matter the coming autumn would probably be a good time to concert measures. So much unfounded speculation has been going on in 'quartz' that even should the mines prove rich there will be much disappointment in regard to dividends, while in the event of the practical working test which will have been made during the summer turning out unfavourable there must be a collapse. At Cariboo at present one would have the ungrateful task of endeavouring to bring down the expectations of the people to the level of common Sense, while definite statements of the value of leads could only be given on the result of assays not Satisfactorally made on the spot. By the Autumn whatever the issue I think they will be ready to be grateful for any assistance.

I am sorry to date still from Victoria, but have been delayed here by circumstances beyond control. Capt Douglas was not here on my arrival. After waiting till the date fixed I engaged another schooner, 'the Wanderer' & have since heard from Douglas that his Craft has been ashore up near Comox & has been considerably damaged. The Wanderer was the only suitable Craft here, & will I think do very well. I have had to wait however til a new suit of Sails could be got ready, besides having to make various little alterations & arrangements on board. We are about ready at last, & tomorrow Morning I hope we shall get away.

P.S. Mr A.C. Anderson of this place would be glad to have one or two additional copies of his notes on N.W. America, reprinted from the Naturalist.[31] I do not know whether there are yet any spare copies. If there are plenty one might also be sent San Francisco.

[Dawson's voyage to the Queen Charlotte Islands is documented in his journal, 'Queen Charlotte Islands Cruise George M. Dawson 1878,' which spans the period 27 May to 17 October 1878. A list of expenditures, several sketches, a letter to Dawson, and some rough notes are excluded.]

May 27 1878. After two weeks of preparation & vexatious delays, get off at 7 P.m. this evening on our Northern cruise. Morning spent in packing up & stowing away things not again required till return to civilization & getting stuff put on board schooner. The sails, which have been the chief part of the delay, ready at last, – the mainsail & jib bent on in the Morning, the foresail after receiving its last stitch brought down & bent in the afternoon. Paid bills & said goodbye for the second or third time to acquaintances in the streets, & now almost myself surprised to find that we are really off

Beat out of Victoria harbour as the light fades from the hills touching with a rosy tint the Summits of the Olympian mountains long after the last glow has gone from the hills about Victoria itself. Pass outside Ogden Pt with a freshening south easterly breeze, & round Trial Island with the last of daylight. Pile away the miscellaneous mass of luggage with which our little Cabin is filled & turn in, tired enough & ready for a good sleep.

May 28. Wind through the night unsteady & light So that <when> on coming on deck about 7 a.m. find we were not much beyond Sydney Island. About 9.15 while engaged stowing away *ictus*[32] shock, followed by a grating sound brought us to the deck, to find that by trying to shove too close to Portland Island we have got on the reefs running off it. The tide fortunately making, so that we get off again at 11 A.m. uninjured & only sorry for the delay. Collected a few star fish & Shells while ashore, the former with many rays like the *Solaster*[33] of the British Coast but larger. The 'tangle' now beginning to reach a considerable length, & floating out with the tide in dense masses from all the reefs & shoal patches. Each stalk tapering downward to its attachment, ending on the surface of the water in a buoy from which a tuft of brown Streamers depend.[34] Wind light, with showers of rain till evening, when a fresh breeze, with occasional spitting rain carried us against the tide through the False Narrows & into Nanaimo Harbour, where, though without any important business of my own I had promised to call to convenience Sabeson.

The vessels lying at the wharves, with the occasional rattle of a truck of coal descending into their holds as they are loaded makes us realize that we are in a Coal-bearing region, where one of natures old store houses is being ransacked for the benefit of the present generation.

The beautiful little islands among which we have been sailing all day, are peculiar in the wall like (mural) cliffs which the outcropping Edges of the Coal bearing Sandstones present on one Side, while the other slopes more gradually down to the water & is generally covered with an open growth of pine <small> trees with occasional little patches of prairie & grass-grown points. There is, however, with all very little soil on these islands & few of them are suited for anything but the maintenance of a few sheep or Cattle. Here & there a settlers house may be seen on some spot of good land, also an occasional little establishment for the trying of Dog-fish[35] & many a little Shanty & potato patch of the Indians, whose canoe may be Seen with a little bag like sail shooting before the wind from one island to another.

The general outlines of these islands show singularly well the dependence of physical features on geology or rather on those arrangements of the rocks which it is the province of geology to study. The sandstones & conglomerates may be traced in long chains of islands following the stricke while from the Shales & softer beds the channels have been hollowed, as has been worked out in detail by Mr. Richardson.[36] To what extent the Strait of Georgia glacier may have assisted in the shaping of the Surface & how far the action of the other denuding agents may have been active it is impossible now to say.[37]

May 29. Obliged much against my wish to remain at Nanaimo till Sabiston had got through business &c. &c. this being his home.[38] Got away at 12.30 P.m. Crept round with a very gentle air by the Channel to Departure bay, & getting outside ran more briskly up the open gulf going east of the Ballinacs.[39] Sky overcast at sunset but a clear spot & bright glow in the North-west.

G.M. DAWSON TO J.W. DAWSON,[40] 29 MAY 1878, NANAIMO

We have got so far on our way northward & hope within a few days to begin work on the Queen Charlotte Islands. The Schooner *Wanderer* seems Suitable for our purpose, with plenty beam & built originally for a pilot-boat. Exactly when it may be possible to write again it is impossible to say but no opportunity wil be neglected of letting you Know at home how we are Please, however, do not let long want of letters Cause any anxiety, as this is only what may be expected. The only vessel Calling regularly at the Northern Ports is the *Otter*[41] & she makes but three trips in the summer, at uncertain dates. Letters addressed to Victoria will be forwarded to Port Simpson near the Mouth of the Skeena River by the otter & if I am lucky I may get one or two mails during the summer. It seems so short a time since we wrote last in Victoria that there is nothing to fill up a longer letter with, so please excuse the present unsatisfactory scrawl & believe me your affectionate son.

May 30. Fair wind having died away during the night, find ourselves at 7 a.m. off Cape Lazo nearly becalmed. Get the tow net rigged & Catch a number of little crustaceans &c. Fair wind gradually springing up Carried us in good style past Mittlenatch Island & to Cape Mudge, but left us off the latter place again nearly becalmed. Favoured by a light air & the tidal current drift past the village of the formerly piratical Ucultas[42] – There appear now to be about 16 houses in all & a large number of Canoes – & on into Seymour narrows. Get through safely with the latter part of the ebb but there being no wind find it difficult to get the vessel out of the stream into Plumper Bay where it was wished

to anchor to wait for the next ebb. Try to bring the schooner Along with sweeps & then more successfully by the boat towing ahead. Get at last into a fair eddy & drift into the bay, anchoring at 9.30 P.m. Went ashore in the boat to a spring the sound of which could be heard from the schooner as it trickled out from the roots of the Cedar trees[43] on the beach. Water beautifully phosphorescent when touched by the oars.
<June 1.> {May 31.} Having rather overslept ourselves did not get anchor up till 7 a.m. Weather quite Calm & in endeavouring to get out of Plumper Bay into the force of the ebb stream got involved in eddys & whorls which Carried us back into the bay. Finally by aid of sweeps got out & continued slowly moving along main channel with the stream, being caught every now & then by an eddy & turned round & round several times. Go off in boat to shore nearly opposite deep Water Bay, for fresh water, having heard a stream running in there. Get at length a little wind & Creep on with freshening breeze first with the tide & then against it. Finally, finding that no progress could be made against the current anchored in the lee of an island[44] beyond the mouth of Nodales Channel, for an hour. Went ashore to examine rocks & collected a few plants. Off again but soon in a perfect calm & moving only with the tide.

Scenery very fine in all directions, the mountains rocky & generally with scarcely any Soil yet supporting Great trees, & in some instances thickly wooded from base to summit. Mountains of Vancouver Shore highest & increasing in height, the Prince of Wales & Newcastle ranges[45] with much snow. Other & still higher mountains of the mainland seen from time to time over those of the islands & bear great fields of virgin snow, which in some cases is seen to form immense drifts behind crags & summits. Passed a couple of Indian Canoes today, the first containing besides a number of Indians, women & men, a white man. The Indians from Some one of the northern inlets & would not even now dare to pass the robbers of Cape Mudge but for the presence of a white. One man <paddling> {rowing} with a short pair of oars, the others, men & women, paddling in the old style. Ictus in heaps in bottom of Canoe & faces daubed with ochre & other pigments giving them a peculiarly repulsive appearance. One woman with a broad mark in red ochre on her upper lip, in the place where a mustache ought to be, looked very comical & defiant.

As I write a breeze Sprung up ahead giving us at least a chance to tack, which with the ebb tide will push us along – slowly.

At about 10 P.m. a steamer going north passed us – probably the Otter.

June 1. Anchored early this morning in a cove N. of Helmkin Island[46] to wait for the ebb tide, the wind having died away. off again at 5 a.m. &

beating all day to windward, part of the time against the flood tide which scarcely allowed us to make anything. wind strong westerly & Cold, though cloudy no rain & high barometer. Little to do but read, eat, & walk the deck wishing we could get along a little faster. Remarkable absence of life, scarcely a gull or other water bird, no seals, porpoises or whales. The great depth of the water may in part account for this. The Vancouver shore Still steep – too & very mountainous tree-clad mountains rising in Many places at once from the water's edge to a great elevation, & still bearing some snow. Extensive valleys, or rather deep persistent valleys, but narrow, run in from Salmon River, Adam's River, & Robson Bight of the Chart. These appear to take a general South Easterly direction, which is also that in all probability of the axes of the mountain Ranges.

June 2. Falling calm early this morning & the tide turning against <up> us, anchored in a little rocky bight just east of Beaver Cove. Get off again after a few hours, about 6 a.m. Almost a dead calm but floated on past Alert Bay, where <West Euson, one> West Huson[47] one of the well known traders &c. of this coast lives. He has a few houses, & a wharf, & near him is a rancherie[48] with a number of Indian houses of the usual build. Most of the Natives are now up Knights Inlet Eulachon fishing[49] & potlatching. The Nimpkish Indians[50] have moved over to Huson's place on Cormorant Island, from their old ground near the mouth of the Nimpkish River.

Becalmed & tide setting against us just west of Cormorant Island. Followed out by a canoe from Alert Bay, which proves to contain a white man & an Indian, the former has just arrived at the bay from his place of abode some fifteen miles further Eastward.[51] Brings a letter left for me by the Otter at Alert Bay, one from G. Hamilton[52] of Stuart Lake which has been to Montreal & back again. Two fine deer in the canoe, one of which I purchased at the moderate price of $1.50. They are hunted by dogs on the small islands, being run off into the water & shot Swimming.

Anchor for some hours in the afternoon off point west of Nimpkish, waiting for the tide. Land & take a couple of photographs[53] & spend the rest of the time fretting at our slow progress. Off again at 4 30 beating up against a strong head wind which has blown up since noon.

The Islands now generally low & the land along shores of Vancouver also much lower than before though not so regularly flat & even as that about Comox for instance. High Mountain Ranges Seen up the Nimpkish Valley beyond the large lake[54] still bearing Much snow & probably always carrying some.

Some fine Douglas firs on Cormorant Island & elsewhere about here but prevalent trees the Menzies Spruce & Hemlock, with some

Cedars[55] &c. These attain a fine growth but their dimensions are by no means appreciated till they can be compared with some object of human construction, such as a house or vessel.

June 3. Off Early & floated westward with the tide; there being no wind – beyond the end of Malcolm Island. Entirely becalmed at about 5 miles from the Vancouver shore till nearly 1 P.m. when a light air began to spring up. Had the large dredge over & brought up some sticky green sand & two or three little brittle stars.[56] Tried the small dredge, but did not get bottom owing to the Current. The light wind coming up, stopped dredging & tried to get on our way. Wind freshened in the afternoon, but as always, dead ahead & beating against it past Fort Rupert, & into Goletas Channel. When off Fort Rupert the wind was quite strong. Passed an indian Canoe scudding before it with a little sail & though pitching her ends clear of the water, apparently making good weather. The occupants in high spirits, bore down on us & took in their sail apparently just to have a look at the Craft, then hoisted & flew away again.

The country from Nimpkish up to North End of Island is all moderately low along shore & for a considerable way back. The hills nowhere of great height. Coal rocks may not improbably occupy a considerable area. Feel almost tempted to stay our Slow progress here & get to work studying them out. After the interruption of the masses of Crystalline rocks about Johnstone Straits &c., the channel between Vancouver & the main appears to open out just as southward in the Gulf of Georgia. The islands are low & the whole appearance like a repetition of the Coal basin to the south-East.

Calms & head winds are our luck it would appear & the barometer gives no promise of a change.

June 4. Beating up Goletas Channel all night with very light head winds. Stopped soon after daylight about 5 miles east of Shushartie Bay to get water. R. went ashore & got specimen of the rock. A fair wind of short duration Carried us to Mouth of Bate Passage. Boarded by a Canoe with a couple of Indians & a boy, which Came off from a little cove, where an Indian House. One of the Indians a chief & calls himself Chip he is said to have saved the lives of several white men in a little vessel near Fitzhugh Sound.[57] The Indians were about to Massacre the crew when he warned them. A number of indians were Killed in the fray which followed. Appear to have no other object in visiting us than to talk & beg a little tobacco.

The tide carrying us through Bate Passage, we Get a flow of wind which after taking us a couple of miles off shore leaves us becalmed till after 2 P.m. rolling gently in the long ground swell of the Pacific agains which there is now no barrier. Porpoises, sea birds, & a sea-lion[58] sporting about in the calm water with patches of Kelp & tangle & floating frayed logs & stumps dipping about in the swell or appearing & disappearing alternately among the long rollers. Drizzling rain for some hours & then a westerly breeze which gradually turned to north, heading us off & died nearly away about dark, while we were yet many miles from Cape Calvert & directly off Cape Caution. At sunset dark neutral clouds in the west & north west Cape Calvert swathed in wisps of fog blue & dark, while southward & eastward the serried peaks of the Coast Range & summits of Vancouver Island, Snow covered, grow with a magnificent rosy hue

Land along shore all round the point of the promontory of Cape Caution, quite low.

June 5. Light breezes all night & fog from about 1 a.m. On coming on deck this morning nearly calm with dense fog & position uncertain. Sight Egg Island during a break in the fog after a time, & also when the mist began finally to clear away about noon saw some of the dangerous rocks of the Sea Otter Group breaking heavily. A long even swell Setting in from the Open Pacific causes a perpetual roar along the exposed shore to leward, & from the outlying rocks great sheets of foam may from time to time be seen to rise. Getting at last a good side wind, run into entrance of Fitzhugh Sound, where then utterly becalmed & remain so till about dark when a light head wind coming down the sound enables us to get under way again. The heavy swell setting in rendered our position most uncomfortable during the delay. The little Craft rolling & tossing too & fro with all her sails booms &c. rattling & flopping in a most irritating manner. The weather has been throughout most adverse to our progress.

Saw a fur seal[59] today. R. had the line out at 60 fathoms for Halibut[60] but unsuccessfully.

The land near the shore about Cape Caution is low as also is that of the island between the two Entrances to River's Inlet,[61] & the southern end & most of western margin of Calvert I. The south end however pretty evidently cryst. & all the rest may not improbably be the same.

A Sea lion is said to have come up quite Close to the vessel last night, & 'bellowed'.

June 6. Nearly becalmed all night, but early in the morning get a light breeze in the right direction. This soon dying away left us again becalmed & drifting out of Fitzhugh Sound with the tide, which at present Seems to have a permanent Set outward. Got the boat ahead towing toward Safety Cove, when a good breeze from the North coming on, ceased towing & began beating in. Breeze died away about sundown to a light air, leaving us again drifting outward with the tide. Run back with the last of the breeze, there being no anchorage along the shore, to

Safety Cove & anchored at ten P.m., thus loozing again nearly all our day's work – a couple of miles or so.

A very fine day, warm & bright.

Two canoes full of Indians from Kitimat[62] *en route* to Victoria passed us while the northerly breeze was blowing strong. Each Canoe with one of the peculiar Sails, careering along before the wind in good style.

Saw a number of whales in the distance this evening breaching repeatedly.[63]

June 7. Morning Calm & cloudy. While waiting for wind got a supply of fresh water & had a glance at the Crystalline rocks of Safety Cove. The tide nearly low & the most beautiful natural acquaria formed by the sides of the cove, which dip steeply down in the clear water. Sea Anemonies of remarkable size & beauty. One variety bright green, a second, with plumose tentacles milk white.[64] Star fishes barnacles & shells of different kinds coating the rocks. A light South Easterly wind springing up got anchor up & beat out of cove. Run for a while with light fair wind, & then met a strong head breeze, against which beating for some time, till it veered westward & made a good side wind for running up the passage kept on all the afternoon making excellent time, & to some extent making up for the terrible delays of the early part of the week.

Fitzhugh Sound & its continuation northward in Fishers Channel, constitute a magnificent water way; wide & free from dangers & straight as an arrow. The land immediately bordering it though hilly, or even in places mountainous is low compared with that at the sides of the higher parts of the inlets penetrating the Coast Range. No high snowy mountains are in sight <at> except at a great distance. Hill & valley alike are densely tree-clad, with Cedar, hemlock, & spruce, spruce hemlock & Cedar in one monotonous spread. The trees do not attain a very great size, & there are many dead trees in the woods, even where no fire has passed. No appearance of any land fit for agriculture, nor of any rocks softer than the old crystalline series.

June 8. Worked up a few miles during the night & this morning in entrance of Lama Passage drifting in with the tide. Air very light all morning but continued to progress slowly, aided by the tide. At noon got a South Easterly breeze with rain which Carried us into Bella Bella (McLochlin Bay of Charts) Found Mr Mckay[65] had gone to Fort Simpson, so unable to present my letters of introduction or see the lode Col Houghton[66] was anxious I should visit. Schooner did not anchor but after I had completed a short visit to the H.B. Post, & given the gentleman in charge a couple of letters for the *Otter* on her down trip, filled & stood on with a light but fair breeze, anchoring in Kynumpt Harbour Millbank Sound.

The H.B. post at Bella Bella[67] prettily situated on a sloping hillside. A small stream coming from a lake behind the Post, falls into the bay near it. A little sloping patch of garden for which most of the soil was, I am told, Carried from some distance. A number of Indian houses & shanties, & a little flotilla of Canoes anchored off. A remarkable target-like white erection on one side of the harbour stated <to be> (on a painted board below it) to be in memory of 'Boston a Bella Bella Chief'.[68] Part of the design on the target a couple of the curious *Coppers*[69] of the Fort Rupert &c. indians.

A very large Canoe here, now lately finished. Said to be 60 feet long & much better finished than that sent to the Centennial.[70] Valued at from $150 to 200 & the Indian who made it expects to be able to sell it to the Fort Rupert Indians for that sum. Of little real use but imposing on State occasions.

Bought a basket of Clams[71] from one of the Bella Bella Indians. Cleaned out some for specimens & had part of the rest in chowder for supper. Visited by a Canoe full of Haida Indians on their way to Bella Bella, & three days from Skidegate. They have besides dried fish & fish oil for trade with these tillicums[72] when they get to Victoria, Some gulls eggs from rocks outside & a young deer. Bought from one of them some fossils which he had been taking to Victoria on Chance of seeing Mr. Richardson there. He had worked for Mr R. when he was on the island. Three more Canoe loads of Haidas are they say also on their way to Victoria not far behind.

Had the line over tonight in about 6 fms where anchored. on hauling up found a huge many rayed star fish which dropped off at Surface disclosing the head & shoulders of a Silver dog fish – a parrot beaked little elasmobranch[73] of remarkable appearance. This fellow had taken the hook & then been Snapped off behind by some larger fish, probably a shark. The star fish promptly appropriated what was left. There would seem to be much activity in the struggle for existence down below.

G.M. DAWSON TO MARGARET DAWSON, 8 JUNE 1878, NEAR BELLA BELLA

Bella Bella is near the Entrance to Millbank Sound, & if we get there this afternoon, as we hope to do, we shall have been nearly twelve days en route from Victoria. I do not know what chance there may be of Sending letters south from here, but as the Otter has probably not yet got so far on her return trip from Wrangel, there may be a pretty early opportunity. We have had a most provoking series of head winds & Calms ever since embarking, & the delay occasioned has been most annoying. This especially as I have been unable to see much of the coast which we have <traversed> passed so slowly, as we have been comparatively seldom at anchor, & then only overnight. We have been drifting about in the passages becalmed, carried by the tides now one way now another.

I do not know that I shall make any stay at Bella Bella, at any rate it will not be for long. A Certain Mr McKay of the Hudson Bay Co. has some lode here which he is prospecting & which I may try to take time to visit.[74] On leaving Bella Bella we shall I think strike across from the mouth of Millbank Sound to the Queen Charlotte Islands & it will not I hope be many days now before we are actually at work. Had I known so much time was to be wasted on the passage I think I should have contented myself with coming to the North End of Vancouver Id. where there appears to be an interesting field. How long we may stay at the Queen Charlottes must depend on the interest of the country there. I think it will be best not to devote much time to the examination of the Crystalline rocks, except in such Cases as something of peculiar interest occurs. To make the main object the definition of the area of Coal measures, & if this proves not very great to leave some time in the Autumn for the north end of Vancouver.

The schooner which I have is Called as I daresay you have been informed before both by myself & Rankine, the 'Wanderer.' She appears well suited for the work, though much time might be saved by having steam power. The 'Captain' is a young <may> man Called Sabiston, the son of a pilot at Nanaimo. He is efficient enough though perhaps a little opinionative. I have besides a man Called Williams,[75] who has sailed often about this part of the Coast, & also about the Queen Charlotte Islands. He is a good worker & will I think do well. Lastly there is the Cook & general utility Man – though also a sailor – A german or something of that sort who goes by the name of 'Dutch Charley'[76] As a Cook he does well & is obliging & civil enough. We have had an abominably lazy time of it so far with seemingly nothing to do but wait for next meal time to come round. The little cabin is littered with books &c. &c. & if only on a pleasure cruise one would be able to get on well enough. The annoyance of loosing time however constantly presses on one, & renders it even difficult to apply oneself to reading.

When off Alert Bay, not far South of the north end of Vancouver, a man came off in a boat with a letter for me, which had arrived the day before by the *Otter*, which left Victoria for the north a couple of days after us. This proved to be a note from Mr Hamilton of Stuart Lake – in the interior – It had been to Montreal & was directed back to Victoria & so up here. We had a few letters & papers before leaving Victoria, but it is now <scarcely> {hardly} likely that we shall hear again for a long time, perhaps a couple of months. The same applies to letters written home, of which this may be the last for some time. I hope you will not allow yourself to feel anxious as it may be quite impossible to let you hear till we return from the Islands. I fear also that there may be some difficulty in getting Rankine back in good time, but I will of course do the best I Can in the matter. This Northern Coast is so much further off in reality, when one comes to travel to it by the slow means which exist, than it appears on the Map. We are drifting down Lama Passage as I write, with the tide but without a breath of wind & when the flood comes in if still Calm there is nothing for it but to anchor somewhere & wait till wind or tide favours us again.

I believe I sent a little memorandum about clothes to be ordered from Edinburgh. These may of course come out any time, with Father's if he orders any.

I cannot think of anything else worth writing about, but if occasion offers will add a postscript at Bella Bella. William[77] will now be with you I suppose. Please give him my love, as also to Anna & all at home. I forgot to take any of William's printed Certificates with me but so far have had no use for them.[78] You might mail one or two to Victoria in case anything should turn up when I get back.

Please tell Father I got a copy of *Nature* before leaving Victoria, also an *Illustrated News*

June 9. off from Kynumpt Harbour early, first with a very light breeze, which, freshening Carried us at last round the rocky islands off Cape Day,[79] & out into the wide Pacific. The long swell breaking furiously on the rocks as we pass them. Six or eight miles off the land the wind going down left us rolling & making a little progress at intervals, westward during the night.

When inside Cape Day Saw something waving on the surface of the water which at first taken for a Shark's fin but proved to be a deer swimming from the South to the north shore of the sound. Williams & R. jumping into the boat dashed after it & succeeding in turning it before it reached the south shore – to which it endeavoured to return – drove it off till bearing down on it with the schooner, I shot it from the deck. A young doe not fully grown but in fair Condition.

June 10. rolling miserably in the swell without wind for some time. Wind then Rising got off westward & <made> {saw} the land near Cape St James before night. Wind freshening went on the other tack to get to windward of the Cape. Blew hard all night rising to a gale with a very heavy sea before morning. Found us about thirty miles northward. Up during much of the night scanning the Channel & trying to preserve things in cabin from coming to grief as they flew from side to side.

June 11. Went round on the other tack <about> & early this morning expecting to make Cape St James. Weather very heavy, but wind gradually failed till land well in sight ahead it left us. Rolling & tossing with out Any wind All the afternoon, in a heavy pitching Sea. Had the Holibut line over but could get no bottom with 90 fms. Breeze from westward springing up about 8 P.m. got under way toward the land.

June 12. Up early this morning to see Cape St James & the southern end of the Queen Charlotte Islands. Took a sketch with bearings. Wind fell very light before we made the point at the southern side of the bay leading in to Houston Stuart Channel. Beating slowly in all morning. Afternoon about 3 P.m. Anchored in snug bight behind Ellen I.[80] of the plan. Took a couple of photographs[81] & made cursory examination of rocks of vicinity.

Weather remarkably fine, warm & summer-like but provokingly little wind for sailing.

Where we are anchored in a snug little bay, rocky islets thickly tree clad down to the shore, with the wooded mountains of the N side of the Channel make a picturesque scene. An Indian house on the shore but has evidently not been inhabited for some time. See no sign of Indians. Rowed round to bay in which *Village* marked on the plan, in the evening, but found only the marks of some old houses.[82]

The appearance of the land about Cape St James very remarkable. Mountains, falling southward toward the cape, & often fronting the sea in bold Cliffs. The little chain of islets[83] off the cape are vertical faced, with rounded tops, bare of trees & apparently the secure resorts of sea fowl. Even the smaller rocks of this group have the same remarkable post like form. Noted a natural archway in the rocks of a promontory a short distance South of Houston Stuart entrance.[84] Another small group of bare whitish sea washed rocks lie some miles offshore north of the Entrance – the Danger rocks[85] of the Chart.

June 13. Breakfast at 6 a.m. & off with Williams Charley & R. in the boat. Explored the southern shore of the channel & Bay as far as Outer Pt.[86] Rain set in shortly after we got away & continued with little intermission throughout the day, Soaking us & making it disagreeable. Lunched at a cove a short distance inside outer Pt. & returned in heavy rain. Stopped at several gull populated rocks & deprived them of their eggs, which – those of them which were not half hatched – made an agreeable addition to our supper. Saw many Seals, a few porpoises, some eagles & innumerable little black & white Guillamettes (?) & a few pairs of a black bird with long bright red bill.[87] No Indians appear, nor have we met with any recent signs of their habitation, which is at least odd. The very abundance of gulls eggs on rocks easily accessible, would seem to urge their prolonged absence.

The rocks everywhere about this passage are crusted with Acorn shells & the large mussels, between tide marks, with occasional patches of Lepas (?) &c. Below high water mark in some places the large urchins are very thickly strewn over the bottom. Sea anemones, starfish[88] &c. &c. are everywhere abundant.

The Mountains & hills evenwhere rise steeply from the shore & the appears to be no arable land. Scarcely indeed any Soil properly so called anywhere. The trees, – among which there appears much dead wood – grasp the almost naked rocks.

June 14. Off early. Rowed westward up the channel against a strong tide. Examined the shores of Rose Harbour,[89] stopping for lunch on its western entrance pt. Continued out to Fanny Pt,[90] without seeing any Indians, but when sailing back, saw a smoke as if made to attract notice, on Anthony I.[91] Shortly after supper a canoe full of Indians Came along side, they having observed our sail, & as we had supposed <seen> made the smoke. All young men, several of them just returned from Victoria with their earnings, & as they informed us 'lots of whiskey' They are having a grand dance today over at the Ranch.[92] The Indian lads well dressed, mostly in civilized costume, & brought with them in their canoe a couple of telescopes! Inform us that they have plenty Holibut & plenty fur seal Skins, the latter they wish to trade. Tried to get all the information I could about the country from them but not very Successfully.

Found fossils in the limestone today. Day cool & showery throughout with South-westerly wind.

June 15. off dredging all morning in the strait opposite our harbour. Drift down with the tide which running strongly, with the dredge rope over the bow. Bottom chiefly shelly & very clean. Get a number of interesting things though much dead stuff. Many beautiful byezoons, some Corals, & one species of brachiopod. (Terebretella?)[93] After lunch set out to look for fossils about two miles westward, opposite the Mouth of Rose Harbour. Efforts Crowned with unexpected success, finding *belemnites ammonites*[94] &c. Sufficient at least to fix the age of the series of rocks which have been examining as Mesozoic & enable them to be correlates with those of other localities.

Morning very fine, but becoming overcast & finally clouding over with occasional showers & becoming cold. Got sextant ready to take latitude at Noon but no chance, & must now leave without getting it. A canoe with two men, a woman & a boy came in tonight. They are I believe from Gold Harbour.[95] They offered to catch us some fish, but on returning the line which they had borrowed for the purpose brought only three sculpins,[96] & three other very small fish. They are going tonight to make a fire in some woods resorted to by sea-birds & club them as they fly past, disturbed in their slumbers.

June 16. Got anchor up & schooner under weigh & then followed the shore, examining it & making running Survey round into next inlet.[97] The schooner had come to an anchor in a snug cove on the south side[98] before we Caught up to her, shortly after noon. Being Sunday, decided not to do any more work today, this especially as I have a headache owing to loss of sleep from mosquitoes last night. Day fine & the swell on the outer shore not too much to prevent Easy landing.

June 17. Heavy rain in the night, & southerly wind. Morning still overcast & showery. Made a rather late start, & occupied till after 4 Pm in making a running survey & examination of the shores of this as yet to me nameless inlet. Rocks uninteresting, & the day on the whole not pleasant, a surf rolling in making landing on the outer points difficult. Saw great numbers of seals today playing in the water & even on the rocks. Some mothers carrying their young on their backs, the two heads coming up out of the water together in a most amusing way.

June 18. A dull threatening morning which soon fulfilled its promise by beginning to rain. Rain & wind in squalls, with low clouds & flying scud on the mountains all day, the monotonous patter still continuing as we swing too & fro with the wind uneasily at anchor. Worked round the coast from last anchorage to Harriet harbour of Pool,[99] in Skincuttle Inlet. Some difficulty in landing on the exposed outer points. The schooner sailing round in the meantime met us at 1 P.m. & after seeing her Safely into the harbour returned to finish work up to mouth of harbour & examine Iron ore deposit marked on Poole's Map.[100]

The general aspect of the Inlet south of this, which we have Just left, & the Country surrounding it is much like that about Houston Stewart Channel. Thickly wooded mountains rise everywhere from the waters Edge to heights frequently exceeding 1000 feet but rarely if ever more than 2000 (eye est.) The shore is generally rocky & the water off it bold. Beaches are unfrequent & not extensive <'Sabiston' harbour> {South} Cove of the plan is a good anchorage for small schooners. Depth of a considerable part of the bight not over about 10 fath. & anchorage at 6 fathoms with good holding ground. The upper end of the inlet is well sheltered, & receives two large streams, but is encumbered with rocky islands & many rocks, rendering access difficult. The timber in this region being of small stature is not of any great prospective value, & <the> agricultural land does not exist.

As we felt our way from point to point round the coast today, in the rain & drizzling mist we continually looked awkwardly for the view round the 'next point'. When the promontory was rounded which gave us a view of the magnificent sheet of water at the entrance to this inlet, its dimensions magnified by the mist appeared grand. We knew not where to look for a harbour but by good luck got into this one.

June 19. Heavy rain during the night, & morning opened with a steady downpour & light southerly wind. Delayed starting out for work, for some time, but at last tempted to go by an appearance of clearing up. Worked along a few miles of coast under great difficulties, the rain recommencing almost immediately after our departure & continuing very heavy, with masses of mist which prevented anything but the land in the immediate vicinity from being seen. Decided to give up survey for today, & got back to Schooner chilled & wet through, for very little. Wind which began to rise about noon soon increased to a gale, which has since continued coming in very heavy squalls over the mountains which hem in our little harbour. Rain leaking through the Cabin roof renders our abode far from comfortable. Reading & attending to other 'home' work during afternoon.

The Schooner has gradually dragged from her first position, under the influence of the gale, to a place nearly in the throat of the harbour, & though both anchors are now out, with plenty cable, she swings uneasily as the squalls strike her, & leave us not without fear that <....> she may drag outside altogether & force us to take to the open. Some of the squalls actually carry the Crests from the little waves in this harbour & scatter them before in a cloud of spray. The holding ground cannot be good, & is probably a fine sandy gravel of granitic fragments, like that composing the little beach near us.

June 20. Continued examination & running survey of coast westward up the Inlet. Finding the entrance of a large bay,[101] – as it proved – though it looked at first a possible passage – obliged to go far enough to prove its character, which took up much time. Came upon a rock around which the tide was rising, quite covered with seals. These on our approach, to the number of 20 or 30 Shuffled off rapidly into the water. Soon they appeared again, heads bobbing up in all directions to get a look at us & then sinking again.

A showery disagreeable day.

June 21. Ran with the wind to the outer Islands at the Mouth of the inlet,[102] & examined the group inwards. Then sailed across to point at N entrance of bay[103] & continued examination of coast westward. Found the abandoned copper mine[104] which Poole Superintended years ago. Little sign now that human beings ever inhabited the spot

Visited a couple of rocks *en route* today from which a few gulls eggs – very acceptable at supper – were obtained. Williams Caught on a rock a young seal just born, with the placenta still lying near it. The little fellow is quite active & seems well able to take care of himself though I fear we have no food suitable for him.

Got back to Schooner after a long run against a head wind across the bay, at 6.15 P.m. A day of rain & cold wind, heavy heavier heaviest being the three categories of the former. The climate indeed seems to be a wet one. This evening as I write it still rains, the drops flying in columns before a strong Southerly wind.

June 22. Heavy rain & wind during the night, & when called at 6 a.m. for breakfast rain descending in sheets. This continued without intermission till 3.30 in the afternoon. No prospect of being able to do any out door work so plotted yesterday's survey & attended to other little matters. About 3.30 the weather suddenly cleared, blue sky appearing. Set to work to make a plan of the harbour, which before 7.30 Pm almost completed. Intended to get a photograph, but clouds & mist hanging about the mountains prevented. Showers are again falling this evening, but not without lucid intervals.

The seal brought in yesterday has been sprawling about the deck all day, uttering from time to time its peculiar Cry of *wah* or *mwah* (a sort of gurgling watery *ma*) in a plaintive tone. At times it becomes quite vociferous for a moment or two & then relapses into silence. At first it shrunk from being handled, but is now becoming quite used to it & even seems to enjoy being Caressed. It already has its likes & dislikes, knowing where to find a warm sheltered corner & how to crawl to it. Williams tied a halibut line to it this morning & threw it over for a swim. It did not like the operation, & when it was about to be repeated in the afternoon, & the rope was brought out it immediately seemed to realize the position & set up a great noise. Thrown into the water it swam well but tried always to regain the deck by attempting to climb the side & continued <uttering> bellowing & mwawing till again hauled in, when it was quite Content. For an animal but one day old it certainly shows great intelligence. It sprawls up to ones legs & pressing its nose about on ones boots or trousers seems to smell them Carefully. It even appears to follow one when one goes away. Its cry reminds one of that which a half drowned baby of goodly size might utter. *Charley* is trying to feed it on oatmeal & water rice & water &c. We have no means of weighing it, but it must be over 30 pounds, or at least up to that figure.

The red billed <plover(?)> {oyster Catcher} haunts the coast here everywhere & one seldom approaches a rocky point without hearing their Cries of alarm as they <hear> see one approaching their nests. They are most grotesque birds with their black plumage, heavy bills & clumsy feet, & withal silly, for the anxious noise they make is a certain means of attracting one to their nests. These are built on bare rocks or rocky points, or gravelly spots of islands &c. They are sometimes on the edge of grassy patches, but never among the grass & no attempt at concealment is made. The eggs two in number are deposited either on some Crumbled portion of the rock where they Cannot roll off, or in a shallow nest, if such it may be called, formed of some small rocky fragments collected together, or of broken & rounded pieces of shell from the beach. In some Cases the nest is conspicuous from being entirely composed of shelly fragments. The birds themselves, though evidently paired generally go in little flocks, feeding together on the shore at low tide. When disturbed they set up a sharp cherraping which they continue not only when on the ground, but when on the wing. Even at night when near their citadels one can now & then hear several of them in conversation, as though they had been awakened by Some disturbance. Their flight is rapid & undulating & when at rest on a rock they frequently sit closely down on the tangle, doubling their legs under them.

June 23. Up at usual time & after breakfast got Anchor up & schooner away for a little bight.[105] about 6 miles off near the entrance to Burnaby Strait. Set to work in boat finishing plan of Harriet Harbour. Next went across to point of Bolkus Is. & made examination & route survey of these. Showers beginning in the morning continued to occur with greater frequency & the day soon became an inquestionably wet one drenching showers with squalls following each other up out of the south-west with scarcely any intermission. Soon thoroughly soaked, chilled, & disgusted & glad to reach the schooner, which we did about 1.30 P.m. Afternoon terribly wet & showers still frequent & very heavy. Mountains all swathed in mist & the wind every now & then rising in force, begins to howl in the rigging. A good fire makes the cabin endurable, but even now difficult to get clothes dry or keep up a sufficient sequence of dry boots & socks. The continuance of this wet weather now begins to become very irksome, interfering as it does so much with our work.

July 24.[106] Morning cloudy as usual, but with high barometer & appearance of clearing, & promise of a better day. Set off notwithstanding light showers to explore Burnaby Strait, & found it opening through into a wide expanse with many deep bays & islands. Got out far enough before turning to satisfy ourselves that an opening probably exists through to the west, & another round to the east, & that it will be safe & desirable to bring the schooner through tomorrow to same anchorage as a base for the exploration of the new region. About 1 P.m. the showers which for some time had been increasing in frequency & duration coalesced into steady rain, which continued, growing heavier all the time, with strong south westerly wind during the rest of the day. About ten miles from the schooner when we turned & had a hard tussle against the wind & driving rain all the way back, getting in after 7 P.m. cold & wet to the skin. Scarcely possible often to see one point from another in taking bearings, & almost impossible to form any correct estimate of distance, or to examine the rocks properly note book sopping all day.

July 25. Heavy rain Continues nearly all night, & on awakening this morning hear the patter still continuous on the deck. Rain continues descending in an uninterrupted deluge all day with heavy squalls of wind, rendering outdoor work impossible & rendering it advisable not to move the schooner from her present shelter in 'Tangle cove'

Tried fishing, but with poor Success, getting only a couple of Sculpins & two Crabs, the latter however of an edible size.

Tangle cove. Is a good anchorage for a small schooner, well sheltered from winds & not too deep. The centre of the entrance between the islands is however occupied by a rock which dries at low water, & must be carefully avoided.

Harriet Harbour is good, even for large vessels, which should enter at the West entrance, keeping nearer to the west shore than to that of Harriet Island, from which shoal water & rocks run out some distance.

The Narrows of Burnaby Strait of Poole may be called *Dolomite Narrows*. They are partly blocked by rocks but may probably be passed with safety by small schooners. The openings running westward to the South & North sides of the Narrows are probably both good harbours,[107] though no soundings were made. The latter especially is very roomy & well sheltered, & might accommodate a large fleet.

> [verso] In passing through Dolomite narrows afterwards find the channel both narrow & crooked with only six to eight feet of water at low tide, probably less at springs. Tidal Current not very strong. [end verso]

All the waters about this end of Burnaby I. should however be navigated with great caution as there are many rocks scattered about, a large proportion of them covered at H.W.

High land & steep slopes of mountains here as elsewhere rise almost always from the shores. The surface is generally covered with trees, a considerable proportion of which are dead at the tops & many scrubby & ill grown. Where a little flat land exists, as at the heads of some of the bays, along the E. side of Burnaby Strait north of the narrows &c. timber of good growth occurs, & does not appear to hold much dead wood. The trees are chiefly spruce (a. menziesii) & hemlock, with some cedar, & alder[108] in groves, especially in the immediate vicinity of the shores. If the climate admits of agriculture these strips of flat land (including probably also some flat islands like Bolkus I) are only suitable & these could only be cleared at great expense of labour. In the narrow passages the trees seem almost rooted in the beach & their branches hang down thickly nearly to the high-water line. Verdant green grass also spreads down literally in some places to meet the tangle growing on the sea washed part of the shore, & indeed judging from our experience so far there is really no reason why the tangle if it can grow without salt should not grow in the woods & on the mountains. Ferns, – the common polypodium[109] – grow abundantly on the trunks & even on the boughs of trees both living & dead, & green moss forms great club-like Masses here & there. Some large trunks overthrown & dead, though scarcely sheltered by other trees bear a perfect garden of moss young trees & bushes though far above the ground

The Mountains near the head of Tangle Cove, & running northward & north westward from it are the highest we have yet seen. Some probably reaching 3000'. Parts of their slopes are bare of trees, & apparently mossy.

June 26. Morning threatening with heavy showers & violent squalls. Got away with schooner after breakfast & sailed down to narrows, hoping to pass without any difficulty. Anchored just before the narrowest part & went ahead in the boat to examine it. Found water very shoal & many rocks & as tide falling & current against <the> us judged prudent to wait till the water deeper & on the rise. Afternoon did some dredging but did not find any very productive bottom. Caught a great number of Crabs, with a hoop rigged with netting baited & put overboard. Got a photo.[110] of the narrows between showers at about 5 P.m. Tried to take latitude at noon, but clouds interfered. Many & heavy showers with squalls all day but some patches of blue sky now & then & better appearance of Clearing than for some time back.

June 27. Heavy rains in the night & early morning, rendering the character of the day so doubtful that breakfast was not ready till 7 O'C. Showers still continue, but barometer rising, & appearance of clearing. Take provisions for two days & blankets, thinking it probable we may get so far from the schooner that it may be best not to return. Measure a base with M.T.[111] & Carry running survey & triangulation down the passage. Day broken with occasional heavy showers, but on the whole a great improvement on any for a long time. Camp at 6 Pm on a contracted little gravelly beach between rocks. Boat anchored out in front.

Try to get obsn. on polaris, but though seen at first, by time instrument ready concealed. Turn in at 11 Pm. a fair night but cloudy & with plenty Mosquitoes about.

Passages & channels seem to open out in all directions with innumerable islands, forming a puzzling maze, espcially when only half-seen through misty clouds & rain.

June 28. Up early & off after breakfast. Spend much of morning coasting a great bay which unexpectedly opened out.[112] Stopped at 11.30 & made arrangements to get obsn. on Sun, which fortunately successful. Got also two photographs[113] looking up the Channel. Came back to schooner in the afternoon most of the way under sail making good time. Looked out a place for the schooner to go to at next move, with work for the boat along the north side of the inlet.

Day almost altogether fine, only a few stray drops falling on us at one time. Sun out for considerable intervals. The higher Mountains to the westward still however continue more or less shrouded, & showers of rain are evidently falling among them from time to time. Looking out on the open sea to the east, the day is evidently quite fine, with scattered Cirro stratus clouds & not a drop of rain falling. The area of great precipitation appears to be pretty local & to centre in the western range of hills of the islands.

June 29. Examined & paced a section formerly seen on the N.W. part of Burnaby I,[114] occupying till noon. Lunch, & took a couple of photographs,[115] & then examined North shore of the island, finishing at the outer North East point,[116] from which the Skincuttle Islands[117] visable to the south. A desolate 'World's End' looking spot. Low land spreading out from the bases of the hills covered with open growth of large, but gnarled trees, the trunks of which fork upwards. Gravel beaches, clean washed & with evidence of heavy surf filling the Crevices between the rough rocky substratum of the shore. The rocks low, but much broken & forming a wide zone between the border of the woods & low tide mark. Shoal & rocky still further out as evidenced by the wide *laminarian zone*.[118] A strong south-easterly wind blowing & over half the horizon a limitless view across the ocean. In a cove just behind the point an old Indian house.[119] The beach not good & Cannot imagine what a house Should be built in Such a place for unless as a stopping place when in rounding the cape winds found adverse.

June 30. Got large dredge put in order this morning, & after getting supply of water, – a somewhat difficult matter as the stream flowing into the bight we are anchored in empties into a lagoon & flows out salt upon the shore at low tide – got off for the north side of the great inlet into which Burnaby Strait opens.[120] Met an Indian canoe with three men, two women, & three dogs, the first we have seen since leaving Houston Stewart. They are going to gather eggs on some of the outlying rocks & have come from *Clue's*[121] to the north. They had a dead seal on board, which was apparently supplying food for *the Crowd*.

Got a Cast of the dredge in 70 fathoms, bringing up a few good shells. Intended getting more dredging on the way across, but sounding in the middle got 94 fathoms, no bottom. Having a bad headache & day not being specially Suitable postponed further dredging. Spent some time looking for an anchorage but finally found a good one between some islands, though reaching it with some difficulty with the sweeps owing to adverse tide & wind.

A beautiful little *cottus*[122] of a new Kind brought up on the hook here, but minus its body, a voracious dog fish having bitten it off as it was drawn up through the water.

No heavy rain today, though several showers, & weather on the whole cloudy.

Wanderer Bight[123] Anchorage for small Craft in 8 fathoms pretty well sheltered. Wide tide flats which drop off Suddenly at Edge of low water mark into deep water. Either of the Coves inside wander bight would probably be better anchorages for schooners though the nearer one much less easily accessible.

July 1. A fine day with much clear sky & scarcely a drop of rain. Off at usual hour & occupied till near 7 p.m. in examining the shores & islands of the opening next north of that to which the name of Juan Perez applied.[124] Got two photos.[125] at noon stop. At work till midnight on plotting & notes.

[verso] Microm. Telescope. July. 1. reads by observations on sun. 18" too little. This probably approximately Correct. [end verso]

July 2. Off at usual time, taking blankets, & food for two days. Worked along coast of inlet to opposite 'Observation point' when Crossed over & continued on west side of inlet. Turned off into a deep inlet[126] which presented some appearance of running through to the west, but find it to terminate. Could find no place to Camp but a little rough beach in this inlet, a triangular patch of depression in a shore-line generally of solid rock plunging into deep water. Mountains rising wall-like above it to a height of some 3000' as steep as trees can grow on, but well covered with vegetation. Looking across the inlet scarce half a mile wide bare granite mountains rise to a height of probably between 4000 & 5000 feet, with their upper gorges & shady hollows full of drifted snow fields. Found scarcely room to spread our blankets down among the great boulders of the shore, & stones from above.

The main inlet which we have been following (Juan Perez) with a remarkably direct general course gives off a number of great bays & long arms to the west.[127] These run up among the mountains of a range which follows nearly Parallel to the west. Low at first gradually increases in height & appears to Culminate near the head of the inlet of tonights camp in great massive mountains, bare & rocky, or even with snow on their summits. – by far the highest we have yet seen. This range is no doubt the axial one of this part of the Q.C. Islands. A fine day.

July 3. Much pestered by the Mosquitoes in the night & breakfasted & got off this morning in a perfect storm of black flies. Coasted out of the inlet on the north side & then continued northward to Island No 19,[128] where lunched & got obsn. for lat. A long row home, Part of way against a strong wind. Got back to schooner at 6.30 & found a canoe alongside with the indians we had seen a day or two since on their way to collect eggs. They tell us that the hot spring of which we have heard so Much is on the island near which we are anchored. Visit the locality, & find a number of Sources all perhaps rising from one place, but flowing out among broken rocks at some distance probably from point of issue. Temperature very various according probably to distance from source &c., but warmest too hot to bear comfortably with the hand. Altogether a considerable body of water.[129] Slight smell of S.H.2 & a barely perceptible saline taste. Full of green confervoid growth. large patches of mossy surface near Sources, not overgrown with sal-lal[130] bushes like rest of I. Heat prevents their growth. On stripping off the moss the ground warm everywhere. Very slight whitish incrustation on stones. Indians bathe in a natural, muddy reservoir.

> [verso] The rock of vicinity of hot spring shows no more sign of recent volcanic origen than that of other neighbouring islands. All these in great part of bedded igneous rocks, but old & dripping at high angles, associated with argillites.[131] Near the intrusive hot spring the predominant material is a whitish rock (see specimen) in <which> association with which & Caught up in it blackish hard argillites. In some places argillites intersected by dykes of the material. Many other places among islands where similar circumstances occur. No reason to argue recent volcanic action. [end verso]

July 4. Went round to the hot spring & took photo. of the mossy patch beneath which it rises. Could not get view embracing the pools & this also. Got second photo looking up the inlet.[132] Returned to schooner & set sail, proceeding up the inlet with a fair wind. Corrected some of sketches by bearings as went along. Took several soundings & had the dredge over in 43 fathoms but nearly lost it on a rough rocky bottom in a tideway. Got plotting brought up to date. The inlet seems to open out in various directions as we advance, there being now four openings to large bays or branch inlets or Channels in sight from near here.[133] Anchored in a snug little harbour (Echo Hr) on the SW shore. Entrance narrow & bold, within expands. Grassy beach at head, & little passage running off to north which opens into a wonderfully secluded inner basin, which however for the most Part shallow. This receives a large stream[134] from the rocky & in part snow clad mountains which are piled at the head of the harbour.

Took photo of harbour after anchoring.[135]

July 5. Off in good time, & worked eastward down one side of the inlet to connect with former furthest pt, then westward back on the other. Turned into opening nearly opposite Echo harbour, which at first supposed to be a large bay, but proved to open out in two directions, the main passage trending north & then eastward,[136] & all clear to the open sea! A strong breeze drawing in, & head tide made it difficult to get far. Turned & after opening the inlet well out & ran back under sail.

Saw two indians fishing at a distance today, & continue to observe many signs of recent chopping, & habitation. We must now be near the Clue village.

Echo harbour. Least depth near mouth about 11 fathoms at H.W. within everywhere about 15 fathoms, shoaling gradually near the head at first, & then running steeply up to a flat nearly dry at low water. Well sheltered from all winds & good soft holding ground.

The main passage,[137] outside Echo Harbour continues to carry the flood from the S.E., the Ebb from the N.W. The tides thus draw through & do not run out from both ends to the open. The current must be over 2 knots at times.

A little fine timber on flats here & there, but very little flat ground. Spruce, hemlock, & Cedar. Yellow Cedar[138] fairly abundant in small trees.

July 6. Took blankets this morning, intending to stay out two days. Rounded the point beyond Echo Harbour to the westward[139] & found a large bay,[140] the inlets at the upper end of which run up among the roots of a mass of high jagged & heavily Snow clad Mountains, probably the highest yet seen.[141] The next opening beyond this large bay is that called Crescent Inlet as a provisional name.[142] It is a fiord, some miles in length, but quite narrow and hemmed in by steep wall sided mountains something like those of the Inlet of July 2.[143] At its head this turns round toward the mountains of the bay first referred to, but without reaching them. The mountains on the north side are the highest. One of conical form when viewed from the entrance proves to be a somewhat prolonged ridge running parallel to the inlet, with several peaks. A second, & considerably higher has a triple summit, and slopes very steeply down to the waters edge. This by reason of a coloured patch on it was called provisionally Red Top.[144] Notwithstanding the steep sided character of this inlet there is a good deal of beach around its sides.

Got observation for lat. at noon, and a photo of the Red top Mountain.[145]

On wishing to camp, after having worked some miles north of mouth of last inlet, could find no water. Ran across the channel, about 2 m. to A bay in a large island,[146] where found a little Spring, plenty wood, a good beach, no flies, & altogether a charming camping place. Put tents up under some fallen and half fallen trees of gigantic size, which form a complete screen seaward.

July 7. Worked along the West shore of the Channel to the open sea, which when opened out displayed a perplexing lot of little islands, some lying very far off the coast. Set [Rip.?] marks & ran across a large bay eastward,[147] but the sun being behind the marks could not seem them distinctly enough to read on them. Could therefore not fix positions of outlying islands. At point where marks to be read from, came on rounding it, Suddenly on the Indian village, Called Clue's village.[148] Went ashore & had a talk with the Indians obtaining some information from them. Asked for the chief Capt. Clue.[149] So called & taken up to his house & introduced to him The village consists of perhaps twelve or fourteen of the large houses usual on the coast, & bristles with totem poles carved into Grotesque figures. Some of the houses entered through holes in the bases of the poles, but Clue's by an ordinary door. Descending some steps one is in a rectangular area depressed somewhat below the level of the ground outside, with several broad steps running round it, on which the family goods, bedding &c. placed. In the Centre a square area not boarded in which a bright fire of small logs burns, the smoke passing off through apertures in the roof above. Clue with some of his friends occupied positions on the further side of the fire from the door. Squatting on Clean mats, several women, who however kept in the background.[150] A couple of boxes brought out on which a well educated Siwash[151] asked Self & R in tolerable English, to sit down. These placed near Clue, & the Indian having first asked who was tyie[152] accordes the nearest post to Clue to me. Had a short conversation & then pleading the late hour got off again on our way to the schooner. Our reception by Clue quite a Ceremonial one, for which occasion offered as he was evidently waiting in some state, & all in order to receive a large party of Skidegate Indians [153] who are expected, & are to join in a bee or potlatch the occasion of which the erection of a new house

[verso] Position of Indian villages in Rocky wave lashed spots.

There are about 32 upright totem poles in the village of all ages, heights, & styles. Of houses about sixteen, including one unfinished, though evidently some time under way. Indians appear very comfortable & moderately clean. The Skidegates down today & a great gambling game in progress, a number of little polished sticks being shuffled up in soft cedar bark.[154] A grand dance in prospect for the evening. [end verso]

Met a canoe on the way with fresh halibut & got a fine large fish for a dollar. Got back to schooner after 8 P.m. having made a prodigiously long round today, but without doing a vast deal of solid work.

July 8. Got anchor up & schooner out of harbour, & then set out in boat with Williams & R for the S. shore of the entrance to the channel. At work all day & back at night to the schooner at her new anchorage. Plotting work till late.

July 9. Cross the inlet & work outwards along the shore of the large island opposite.[155] Getting to the Indian village pay a rather lengthened call with the object of getting such facts as I can about possible coal &c. Hear confirmation of the story of a spot on one of the islands outside that of the hot spring, from which bitumen, or something like it oozes.[156] Present Chief Klue with a pound of tobacco, & finding no objection made take a photo.[157] of the village. Would have taken several but the rain threatening all the morning now began. Lunched near the village & then ran across to the outer island[158] on the outer part of which a very good section of great thickness of the dark argillites & flaggy limestones[159] here & there a poorly preserved amonite. Came back amid rain & wind with a heavy breaking sea, round the east end of the island & got on board schooner at six p.m. Heavy rain still continues at 11 P.m. At work plotting <work> notes &c. till late.

July 10. A very wet night, & heavy rain early this morning, rendering one uncertain whether to start along the open coast for Cumshewas,[160] as had intended, or not. This especially as barometer very low & not showing any inclination to rise. All appearance of Clearing <at> after breakfast, got anchor up & away. Called at the Indian village to try to engage an indian known to Charly, one of the Skidegates who are visiting here. Find that all left early this Morning. Got some information from the Indians who Came off in Canoe, by which it appeared that a large inlet between this & Cumshewas, with a large island in its entrance.[161] Wind light, did not get to end of former work till so near noon that thought it best to have lunch on board. Got off shortly after noon, instructing the schooner to go round to the other side of the island & find an anchorage. Rowed about 8 Miles up a long inlet,[162] which finally became very narrow, & though still running on, took such direction that seemed very improbable it would turn Seaward again. In much doubt what to do, as getting late, but finally decided to sail back down the inlet, & look for the schooner on the other side of the supposed island. Did so, but on rounding the point, find a second great inlet,[163] with the wind again blowing out. Caught sight of the sail far up the inlet, & rowed laboriously down to a fine harbour[164] in which schooner at anchor, a distance of probably eight miles arrived on board after 9 P.m. guided by fog horn & lantern. R. tired out with rowing & self cold & tired also.

Williams rocked overboard by the boom near the point had a wet & cold row to the schooner. Fortunately in falling Caught side of boat, or might have been more serious.

Long heavy swell from seaward today & strong winds from the land. – westerly winds.

Passing the Indian vil. this morning in bright sunshine all alive like an anthill with Indians in blue, red Green & white blankets. Hard at work holding the 'bee' for the erection of a new house for Chief Klue. Cedar planks of great size hewn out long since in anticipation, towed ashore some days ago, now being dragged up the beach by the united efforts of the motly gathering, harnessing themselves in clusters to ropes as one sees in old Egyptian pictures of the movement of masses of stone, though numbers engaged here of course smaller. Heaving & howling or ye-hoving in strange tones to encourage themselves as they strain at the drawing. The large Cedar beam lying on the beach is being elaborately carves as a door pillar for Klue's new house.[165]

July 11. Off as usual though morning threatening, to examine the maze of inlets &c. which has opened up between Laskeek & Cumshewas. Drizzling showers in the morning, followed in the afternoon by almost continuous rain with squalls of wind. These seem concentrating into a blow from the South East.

On our return, after joining with yesterdays work at bottom of first inlet, find a heavy wind & sea rolling in which gave us a hard tussle to weather the shore, with its numerous points & get into our harbour. Crept slowly round among sharp chopping seas, & finally, having opened the sea, tilted into the mouth of the harbour on grand rollers too large to be dangerous to the boat, but which sweeping on the rocks at the entrance surrounded them with a seething mass of spray & foam.

Weather still wet, barometer going down & the sound of the breakers outside constant.

July 12. Raining nearly all night & this morning still raining with wind & heavy swell & low barometer. Judged it best not to attempt out door work & consequently devoted time to reading &c. Schooner beached at high tide & bottom scraped in afternoon. Pm only showery Took a short walk along the beach & went for an hour fishing, Catching a few rock Cod.[166] Evening barometer rising & some appearance of clearing though light Showers Still continue.

July 13. Every appearance of clearing this a.m. with high barometer but showers beginning soon became almost continuous & continued very wet till after Noon. Later cleared up partially & ceased raining. Got a rather late start, the boat having grounded owing to the extremely low state of the tide. Did a good day's work however, getting sight of the extremities of the three remaining branches of this great inlet. After completing work got a good breeze to carry us nearly all way back to schooner, & having arrived early at the Mouth of the harbour put a line over to try for a fish. Found the fishing very good & caught about 3 dozen good sized 'rock cod' of at least four species. Very gaily coloured & spawning to a degree. First rate eating however. At work till late plotting & writing up notes. Got observation on polaris after 11 P.m.

July 14. A rather late start again owing to Charley who overslept himself. Worked round the shore in the boat nearly to the Indian village (Skedan)[167] at the S entrance to Cumshewa's Harbour. Then boarded the schooner which not far behind & run on with very light wind anchoring on N side of bay at about 10 P.m. Morning very bright & fine but afternoon clouding & evening showery & calm. A couple of Canoes Came out from the Indian village to meet us, freighted with Indians who brought a quantity of wooden bowls &c. to trade as curiosities. The Chief Skedan[168] in the larger Canoe, dressed in a good suit of black, a middle aged man of less power than Klue apparently & commonplace mind. Presented his 'papers'[169] which simply said that he was a good sort of Indian &c. &c., with the exception of one which written by one of a number of people who were shipwrecked in 1852-53 on the coast, in the schooner <name> {Olympia}.[170] The writer said he had no doubt their lives due to this man & another who objected to the other Indians carrying out their intention of murdering them at once. Gave the Chief a small present of tobacco & proceeded to bargain for some of the Indian curiosities. Skedan says very few indians here at present, nearly all in 'Vic-toi'[171] village occupies an exposed situation on a gravelly neck at the point South of Cumshewas Hr. It bristles with 'posts' & must be more carefully examined.

July 15. Raining with thick mist & strong wind this morning, a state of affairs which continued with little abatement during day. A heavy sea running in the bay & breaking on the shore. Remained about schooner all day, making only a short extension to the shore to see the rocks, which diorites. The little cove on the North Side of Cumshewas Hr, in which we are anchored is that in which McCoy built a house some years since for trading purposes.[172] Hoped to induce the Indians to catch dog-fish & make oil but found it did not pay, the Indians constantly going to Victoria & being without habits of steady industry. House Sill Standing. The Cumshewa Indian Village about 1$\frac{1}{2}$ mile further up the bay on the same side. Clearly visable from here, with its row of 'posts'

& houses. Very few Indians said to be there at present.

Chief Skedan finding it a profitable business yesterday came across today with a lot more masks, rattles &c. &c. to sell & succeeded in inducing me to take a good many at rather exhorbitant rates.

Told by the Indians that only three kinds of mammals in the *Haida illihie*,[173] viz: Black bears, now in the mountains, but common along shore when the hook billed Salmon begin to run; Marten (as far as I can make out) & Otter. There is also a mouse, however, or Small mouse-like Animal two of which have been seen by us along the beach,[174] but of the existence of which my Indian informant did not seem to be aware. Says there are plenty frogs, but no snakes on the islands.[175]

July 16. Being unable to learn from the Indians exactly how far it was to the head of Cumshewa's Harbour, or rather inlet, took our blankets with us this morning. Took three photographs of the Indian village *en route*, with special reference to the Curious *totem poles*.[176]

In examining the shore found very interesting sections of the Coal bearing rocks, with abundant fossils. Did in consequence only a comparatively small stretch of Coast, camping in a little well sheltered cove[177] with good water & plenty dry wood. Altogether a charming camp. Up till late getting obsn. of Polaris for lat.

Fine & very warm day.

[verso] General character about Cumshewas
separation between high & low country
shoal character of harbour. [end verso]

July 17. Boat aground this morning but got her off without much trouble & proceeded on our way, examining the head of the Inlet. Returned along south shore, getting back to schooner at 7.30 after a long pull. Took one photo of snowy mountains near head of inlet.[178] Tried to get Sun. lat. at noon but cloudy. – A fine day though more overcast, & not So warm as yesterday. On returning on board found three Indian women with little things for sale among which two large spoons made from horns of the Mountain Sheep[179] very ingeniously. Bought one for $1.00 though $2.00 at first asked for it.

July 18. Besieged by Indians with various things to sell this morning, curiosities new potatoes about the size of wallnuts[180] &c. Got away at last, & pulling across the harbour Carried work on to Skedan's village, arriving there about noon. Took five photographs of the village & totem posts,[181] which here appear very interesting. Had lunch, & then examined the large bay S. of the village,[182] Connecting Satisfactorelly with the other work. Had hoped to have a fair wind back but this dying away, a wind sprang up Out of the inlet, giving us a long pull back against wind & tide. Arrived on board at 8 P.m. Soon after dark Capt Klue & three of his people arrived, disgusted that they had not been knowing enough to offer *ictus* for Sale when we were {in} their country, & having heard that Skedan was making a big thing out of us. Brought with them one remarkable mask with a nose about 6 feet long, a dancing pole highly prized & gaily painted & a head ornament composed of cedar bark into a ring of which a great number of imitation arrows, in wood & feathered, were stuck All these they valued much & had evidently brought their best things to cut out Skedan & his friends, mistaking our taste for homely illustrative articles for those gaudy & good to the uninstructed Siwash. After a deal of talking felt almost obliged to buy one & got the mast for $2.00.[183] Had also to give them permission to sleep on deck as they showed signs of staying in the cabin all night if not.

Skedan's village shows signs of having passed its best days. Some time since though not quite so deserted as Cumshewa's It has always been a larger village & many of the houses are still inhabited. Most, however, look old & moss grown & the totem poles have the same aspect. Of houses there are about sixteen, of totem poles about 44. These last seem to be put up not merely as hereditary family Crests, but in memory of the dead. An old woman is getting one put up now, at which most of the tribe is away at work, in memory of a daughter who died in Victoria lately. The flat topped, boarded totems are more frequent in this village than elsewhere seen.[184] One of these shows a curious figure leaning forward & holding in its paws a genuine *Copper*[185] like those described to me by Mr Moffat[186] as in great request & much worth among the Ft Rupert indians At least one other *Copper* in view on the posts here, but the second observed not in Evident relation to any of the Carved figures

The village situated as usual in a wild rocky & sea-washed place, on a shingly neck of land connecting a broken little rocky peninsula with the shore. on the peninsula two remarkable, symmetrical, conical hillocks, which form a good landmark when coming from the South.

July 19. Examined rocks between anchorage & Indian village, including the metalliferous veinlets pointed out by the Indians. At the village found the old Chief himself, Cumshewa.[187] He had heard that I wanted to see the reported coal & was ready to come with us & show it, on the understanding that he should be paid. This I promised provided he really showed us some coal. A rather pleasant & quiet old Indian, speaking very little Chinook but trying to make himself as agreeable & useful as possible. Found the coal in small fragments in sandstone. Examined locality & collected specimens. After lunch paced several miles of beach for section, & remained Collecting fossils, which occur in great abundance in Some places – so long that did not get back on board Schooner again till 8 P.m. At work late on notes &c. Found a canoe alongside on return, with a number of Holibut Indians appeared considerably disappointed that I should buy only one, which I did for half a dollar.

July 20. Anchor up Early this morning, & spent greater part of bay dredging outside the harbour. Could find only shelly bottom. Which though yielding no great variety, gave some things of interest. Evening visited the Indian Village, plotting work & notes till late.

About sundown two large Canoes with two masts Each, & the forward one with a large flat hoisted, hove in sight round the point. Turn out to be Kit-Katla Chimseyan Indians with loads of oolachen grease for Sale.[188] They have slept only two nights on the way from Kit-katla They come here on a regular trading expedition, & expect to carry back chiefly blankets in place of their oil. Only a few of the Haidas Seem to understand Chimseyan so that have the curious spectacle of Indians communicating with each other by this vehicle. Quite a picturesque scene when the Canoes grounded & the Kit Katlans assisted by the Haidas carry up blankets used as bedding, miscellains little things & the Cedar bark boxes which hold the precious oil.[189]

Arrival of Chimseyans to sell oolachen grease to Haidas. Evening sky just loosing glow of sunset. Two Canoes appear round point. Sails clued up to masts, of which each Canoe has two. A bright red piece of bunting flying from the Canoe ahead. Who are these. Haida looking attentively pronounced Chinseyans & proved correct. Soon in good view. Greater part of occupants women. All fairly well dressed & wearing Clean blankets to make a good appearance on arrival. Faces of some painted black or dull red giving a wild appearance, which rendered comical by the top pieces which civilized [tilis/titis?] of various patterns but all intended for the male sex. {Some children, several very patently half breeds.} Dip paddles with a slow monotonous persistency after a long day's work. Tell us that have only slept two nights since leaving Kit-Katla. Come in to beach at Haida Village & received by its inhabitants, who appear anxious to assist in every way. Bark boxes holding the greese set into the water beside the canoes. Other things carried carefully ashore. Canoes hauled up, & then the greese boxes carried carefully up beyond high-water mark, the villagers assisting. A large box an inconvenient load for one man. Regular system of merchandise. Expect to get blankets from the Haidas for the grease, each box being worth from 6 to 10 blankets or say from $12 to $20. The grease notwithstanding its offensive odour a very favourite article of diet with the Indians. Remember that also packed into the interior by the *grease trail*[190] & in fact radiates in all directions from the great oolachen fisheries of the northern part of the Coast. Indians in Victoria value very much, but there very dear.

Cumshewa's Harbour of map really a long inlet & should be Called Cumshewa's Inlet. It differs in its somewhat greater width & the low character of the land on its Northern shore from the other inlets to the south, & in fact marks the junction of the Mountainous & flat countries on the East Coast of the island. There is more beach along its shores than in the southern inlets & wider tide flats. These only indication of shoaler water, which not only in the Inlet itself, but now extends far off the coast, probably marking the submarine extension of the soft Coal bearing formation in an uncrumpled state. The heads of Some parts of the Inlet, however, appear deep & have bold shores, this only on a smaller scale what found on grand Scale in Many of larger fjords. The Mountains to south bear snow in abundance which without doubt lasts all Summer. They are as high as any yet seen & the mountainous country does not therefore die away but suddenly breaks down in this direction.

The Southern head of the Inlet almost (quite?) meets <that of> an arm of that explored last.[191] From the southern or south Western extremity an Indian trail leads over to the head of Tasso or Tasoo Harbour,[192] which Can be traversed in half a day, & is not infrequently used by the Indians, who do not permanently reside in Tasoo. <From the Mouth of the stream near McCoy's House, & opposite our anchorage, another trail starts for Gold Harbour.[193] This is further off, the journey occupying two ordinary days travel. Trail said to go through a comparatively low country & to pass by one or more lakes. Is probably pretty easily traversed or Indians would go round by Skidegate by water.>

[verso] {'no such trail'} [end verso]

The two symmetrical little knolls near Skedans Village form good landmark from south for southern entrance of harbour. <Cumshewas> {not Cumshewas} Island[194] on the north bristles like a porcupine with dead trees. Must be dangerous to run into harbour or Inlet in thick or dark weather for besides general shoalness of shores in this vicinity, An extensive rocky reef[195] lies a little to north of centre of opening well outside. Beyond this perhaps half a mile further East is a second reef bare only at low water. Others not seen may exist besides these, & the low Islands to the south[196] are probably well fenced with reefs also.

At southern side of mouth of harbour very regular depth averaging 20 fathoms with shelly bottom everywhere. The bar at the entrance to the harbour is not a spit as shown in Chart but a wide bank or flat stretching from the south shore. Comparatively little of it dries, but at this season all thickly covered with kelp, which buoys it well.[197]

Vessels should be very cautious in approaching the shore everywhere in this vicinity either outside or inside the inlet as it runs shoal a long way. The bay in which McCoy's house is built affords fair anchorage for small schooners off the edge of the tide flat which wide. The anchorage marked, with soundings, inside Island on North Shore[198] must be a mistake, for at high tide can see bottom nearly all way across the supposed anchorage & kelp abundant off its mouth. The best anchorage for large vessels probably on S shore opposite the long beach which fronts the low ground of estuary of a large stream.[199] This probably one of the places marked on the chart.

July 21. A fair day on the whole though generally overcast, & Showery in the evening, with mist. Got off in good time & Carefully examined & paced the section along part of N. shore of harbour. Chief object to get an approximate thickness for the Shaly argillites &c. In this probably fairly Successful. In returning called at Indian Village & engaged a young Indian to whom I had spoken before, to go with us to Skidegate & perhaps further. A rather heavy-looking fellow, whose chief peculiarities appear to be a long back & short legs. Speaks Chinook of Course, but also some English & is too proficient in swearing, a habit no doubt contracted on former cruises on schooners. Evening writing up notes, Sorting specimens &c. till late.

July 22. A dull morning with dropping rain & thick fog. Barometer high however, & as the Indian boy engaged yesterday has turned up, & otherwise all ready. Decide to set out for Skidegate, leaving the Schooner to follow as soon as she Can. The fog gradually Cleared up, but heavy rain soon began & continued without intermission almost all the morning, Soaking us Completely. After lunch became fair, though still clouded. Sea moderately Calm, & So made good progress along the Coast, Camping at about 5.30 Pm in a wide bay.[200] Hauled up boat on skids.

July 23. Off Early, with fair weather & Calm sea soon made Spit Point at entrance to Skidegate & then examined along south shore of inlet for some distance, stopping at the point East of Alliford Bay, in a pretty little cove with sandy beach. The Schooner not having yet appeared, decide to stay here where we have a good view of the entrance to the harbour in preference to going across to the rendevous near the Indian Village.[201] Got a good fire going & soon very comfortable

> [verso] From the mouth of the Cumshewa's Inlet to Skidegate Inlet the coast is all low, rising in a few places at the shore to a height of 200' to 300' only, & generally very much less than this, though gain-

ing some such elevation at no very great distance back.[202] A series of wide open bays, separated by low spit-like points is found. The points are generally elevated about 20 feet above high water & are composed of gravel &c., being evidently rough spits formed by wash in opposing directions when the sea stood somewhat higher than at present. In some places these now bear fine woods. With the low shore – so different to that we have been accustomed to further south – the beach becomes flat, & very shoal water extends a great way off shore. Near Cumshewa's the beaches are almost altogether formed of boulders, but toward Skidegate they become finer, though still plentifully Strewn with erratics, which occasionally become very thick About the projecting points. Large boulders appear far from Shore at low tide. The erratics probably derived from the mts. of the interior of the Island. Spit Point at South Entrance to Skidegate particularly flat, & runs off into the long bar which Stretches across mouth of Inlet. This very shoal seaward, but falls off suddenly inward, into deepwater. The land about the Entrance to the inlet on the North Side is also quite low & flat, & from a little height Can be seen stretching a great way, the horizon line being in the end broken only by Scattered clusters of trees projecting above it. Further on, the hills begin to rise on both sides of the Inlet, & towards its head ranges of snowy mountains appear, probably equaling in height any we have yet seen. [end verso]

July 24. Raining in the Morning, but caught sight of schooner beating into the Inlet soon after daylight. Got breakfast, & then put things in boat & started across toward her. Got on board & soon anchored in bight opposite Woodcocks Fishery,[203] now run by a man called Smith, & his partner Macfarlane?.[204] Found a small sloop belonging to Collins[205] lying at Anchor, but just waiting for a wind to set off, & going direct to Victoria. Wrote letters to forward by this Chance, & after Early lunch examined the coast westward for a considerable distance. Collecting fossils & marking down attitude of rocks. Went round in the evening to the village to see the Indians dance. Returned late & went to bed at once quite tired out.

[verso] *Indian Dance.*[206] Landing from our boat after dark at the south end of the fine sandy beach opposite Skidegate Village, find this part of the town apparently quite deserted, but see some dim light at a distance, & hear the monotonous Sound of the drum <at some distance>. Scrambling as best we may along the path which winds along the front of the row of houses, & narrowly escaping falls on the various obstructions in it, we reach the front of the house in

which the dance is going on. The door is to one side of the middle, & not through the bottom of the totem post as in the older fashioned buildings. Pushing it open a glare of light flashes out, & entering, we find ourselves behind & among the dancers, who stand inside the house with their backs to the front wall. Pushing through them we cross the open space in which the fire, – well supplied with resinous logs is burning, & seat ourselves on the floor among a crowd of onlookers at the further End, having first taken off our hats at the request of some Indians near us. The house oblong of the usual Shape, but not excavated in the Centre as is often the case. The floor boarded, with the exception of a square space of earth in the middle for the fire. The Chattels of the family piled here & there in heaps along the walls, leaving the greater part of the interior Clear. The dancers as already described occupy the front end, the audience the sides & further {in} end of the house. The smoke from the fire, – which the only light – escaping by wide openings in the roof. The Audience nearly fill the building, squatting in various attitudes on the floor, & Consisting of Men women & children of all ages. Their faces all turned forward & expressive of various emotions lit up by the fire. {Skidegate village about 25 houses & some 53 totem posts.} The performers in this instance about twenty in number, dressed according to no uniform plan but got up in their best clothes, or at least their most gaudy ones, with the addition of certain ornaments &c. appropriate to the occasion. All or nearly all wore head dresses, variously constructed of cedar bark rope ornamented with feathers &c. or as in one case with a bristling circle of the whiskers of the Sea-lion. Shoulder girdles made of Cedar-bark rope, variously ornamented & coloured, with tassels &c. very common. One man wore gaiters covered with fringes of strung puffin[207] bills which rattled as he moved. Nearly if not all held sprigs of fresh spruce, & were covered about the head with downy feathers which also filled the warm atmosphere of the house. Rattles were also in order. Different from the rest however, five women who stood in front, dressed with some uniformity, Several having the peculiarly beautiful mountain goat shawls which are purchased from the Mainland Indians.[208] The head-dresses of these women were also pretty nearly the same consisting of Small mask faces Carved in wood & inlaid with haliotis shell,[209] these Attached to Cedar bark & built round with gay feathers &c. stood above the forehead. The faces of the women – as if All engaged in the dance – gaily painted, vermillion being the favourite colour. Another important feature the master of the ceremonies, who stood in the middle of the back row, slightly higher than the rest, not particularly gaily dressed, but holding a long thin stick with which he kept time

& lead off the singing. A second man in the dance also held a stick, somewhat different from the first, being white & with a split & trimmed feather in the top. Do not know whether this man any part of the Ceremony. He, however, had a pronounced place on one side in front of the row of dancers.

The performer on the drum – a flat tambourine-looking article formed of hide stretched on a hoop – Sat opposite the dancers & near the fire, So that they Could mutually see each others movements. The drum beaten very regularly in 'double knocks,' thus – tum tum – tum tum – tum tum – &c!

With this the dancers kept time in a sort of Chant or Song to which words appeared Set, & which rose to a loud pitch or fell lower according to the motions of the Master of the Ceremonies, who besides keeping up the time now & then slips in a few words of direction or exhortation. To the drumming the dancing also keeps time, following it closely. At every beat a spasmodic twitch passes through the crowd of dancers, who scarcely move their feet from the floor but move by {double} jerks, shuffling their feet a little at the same time. Those who dance best, especially the 5 women already alluded to turn about half round in three or four jerks, & then turn back again in the next three or four. The heads of these women also moved as though loose & set on pivots, jerking idiotically as they moved. When the chorus swells to *forte*, the rattles are plied with tenfold vigour & the noise becomes very great. After a performance of ten Minutes or so the Master of Ceremonies gives a sign & all stop, ending with a loud *Hugh*! After a few minutes repose the movement begins again, with the drum.

The crowd of gaily dressed, gaily painted savages by the kind light of the fire present a rather brave & imposing appearance, & when in the heat of the dance I suppose the Indians may yet almost imagine the old palmy days when hundreds Crowded the village & nothing had eclipsed the grandeur of their ceremonies & doings, to remain. The occasion of the dance as far as I could learn, was the passing of a young man one degree toward being a 'Chief,' or head of a family. They gradually take rank passing a stage when they get a house erected, and becoming of some importance when the totem post has been erected a Potlatch of blankets occurs on each occasion. [end verso]

G.M. DAWSON TO J.W. DAWSON, 24 JULY 1878, SKIDEGATE

Having arrived at a little trading post, or rather station for dog-fishing & the manufacture of oil, we find a small sloop on the point of Sailing for Victoria, & take the opportunity of writing a few lines to let you know that we are well & where we are.

No news from home has reached us since leaving Victoria & there is no chance of our hearing anything till we reach Fort Simpson, which may be some time next month.

We have now Completed a running Survey along the East Coast from near the South End of the Islands to this place. The shore has proved very complicated a maze of inlets & islands occupying much time in examining even imperfectly. The geology is interesting enough, though no points of special importance have occurred yet. All the fossils hithertoo obtained are Mesozoic, but appear to be of two horizons. The highest that of the Coal bearing rocks, the lower perhaps Triassic or somewhere in that vicinity. I must not, however, occupy all my time in giving geological notes. – The weather has proved not quite so bad as some accounts would lead one to suppose though quite wet enough at times, & Seldom too warm or dry. of late it has been considerably improved & I hope it may continue so as work goes on so much faster.

I will probably remain here a week, & may be in the vicinity for a longer time should I decide to visit a few places on the West Coast *via* the Skidegate Strait. <?Skidegate Channel?> If ten days or a fortnight brings all this to a conclusion, less than a week should see us round at Masset. Two weeks should see us through about the North End of the Island, & we may then probably make over to Fort Simpson, & set on our way Southward, with time to spend on the N. end of Vancouver should weather admit. One Cannot of course, however count on keeping fixed dates, nor Calculate the time necessary for the southward trip. Should this reach safely it will be by a lucky Chance & no opportunity may again occur of sending a mail till we are nearly back to Vancouver at least.

Rankine is well, & now occupied in writing at some length to Mother, so that you may have more news than contained in this Short note.

July 25. Heavy rain this morning renders us rather late in starting. Crossed the inlet & worked slowly round Alliford Bay & E. end of Maude Island. Collecting a number of fossils. One place where particularly good near the Indian village (Gold Harbour tribe) stopped some time & soon surrounded by nearly all the inhabitants of the place, men women & children. The latter were specially troublesome as they set to work pounding out, or trying to pound out the fossils with stone &c. & no doubt ruined many. Told them on leaving that if they would collect a number & bring to schooner tomorrow, I would pay them. Back for supper a little befor 7 P.m.

A long visit from Smith & his partner.

The climate is said to be exceedingly wet here in winter, – only three fair days last winter – For two last winters no snow, but sometimes a good depth. This summer Said to be more southerly, with clouds & rain than usual, indeed quite exceptionally so according to Smith.

July 26. Off in good time. A remarkably fine bright morning. went round to Indian village & took three photos. of it.[210] Then examined rocks at next point outward, but finding beyond nothing but Sand & beach, returned, looking at Bare & Tree Islands[211] *en route* & rowed on to where left off work yesterday. Worked on round Maude & Lina Islands &c. getting back to Schooner with fair wind about 6.30 P.m. Showers during day notwithstanding the fine appearance of the morning. After supper went with Mr Smith to see some fine spruce trees behind the fishery. These have been used to some extent for making barrels for oil, & a quantity of wood lies ready split up for that purpose. The timber very fine. Tall & straight, though not of such great diameter as occasionally obtained.

July 27. Away in good time, Crossing to the gold harbour indian village to find a Man acquainted with the locality of the Coal said to exist further up the inlet than the mine,[212] who would go with me to the place when I am ready. Found the last two or three canoes just leaving for Gold Harbour, whither the whole tribe is now *en route* to make a fishery of Mackerel[213] for oil, as I understand it. Find two old men who explained as well as they could to Johnny where the Coal was. Worked along Coast south of Maude Island, with fine weather but a strong head wind, which delayed us much. Found one remarkably good locality for fossils on South Island[214] stayed a while to Collect. Got back to schooner, – now Moved up to anchor cove[215] – by 6.15 P.m.

A fine day & evening with rain on the western hills but none eastward.

July 28. Breakfast at 7 a.m. Then sailed to various islands noting the rocks, & returned along shore examining the section from Lina Island. Got back to schooner at 1.30. Dinner. Did not go out again in P.m. reading, going over notes &c. Morning fine, gradually clouding & persistent heavy rain in the afternoon.

Found an Indian grave. on Reef Island, which my Indian informs me is that of a doctor, who died about ten years ago. A square box like structure about 5 feet high, made of cedar boards split out, & roofed with the same, but with the addition of a pile of stones to keep the whole in place. A board having fallen out, looked in to the last home. The back of the house covered within by a neat Cedar bark mat. The body in a sitting posture, the Knees having originally been near the chin, but the whole now slipped down somewhat. A large red blanket wrapped round the shoulders. The hair, still in situ & black & glossy, done up in a knot on top of the head & secured by a couple of Carved bone pins. A carved dancing stick leaning up in one corner, in front of the Knees a square cedar box, no doubt containing other necessary properties for the next world. The tomb under some spreading spruce trees near the rocky edge of the island overlooking the water. Do not know why in this instance they have departed from the usual custom of putting the dead in similar little houses behind the houses of the living in the village.[216]

July 29. Examining the rocks along the shores of the *Long Arm*[217] all day today, getting back about 6.30 after finishing with a couple of hours fatiguing Scramble through the gigantic & dense woods of the country, in search of 'No. 2. Coal Mine'[218] – a search which proved unsuccessful. Day windy & with shower Succeeding shower in almost uninterrupted succession, the Mountains being constantly veiled with mist

Found the *Echinopanax*[219] abundant in the woods today for the first time.

July 30. Had boat put on beach, & Williams set to work to make some repairs on her, Indian to Cut wood for fires, & then set out with R. & Charley for the Coal Mines.[220] Walked up the track, which broad & well laid, with two fine bridges, one of which has now fallen. Trees have also fallen across the track in Many places, & thickets of bushes & weeds grow upon it. The bunkers screens &c. & all the arrangements for shipment have been very complete. The broad gauge line leads from the screens to the wharf, about a mile. The screens at foot of an incline, which worked by fall against empty trucks, double track drum with friction brake at top. From this tramway leads up a pretty steep incline to the Hoopers Creek Tunnel, & into it. Some grading done on line toward Hutchins Tunnel, but rails probably never laid. A rank growth is rapidly covering everything, chiefly Salmon berry bushes, with fire-weed &c.[221] In ten years it will scarcely be possible to trace anything. Tunnels beginning to fall in, & some partly full of water from blocking at mouth Had a fatiguing day of it, Scrambling through the woods & up the terribly tangled & encumbered stream beds. Rain falling steadily, bushes & trees completely saturated, ground generally almost spongy & full of water. Returned about 3 Pm wet to the skin, cold & tired. Plotted some of back work, wrote up notes &c.

Perhaps sufficient certainty may not very unreasonably have been supposed for the Coal seams, but by obtaining thorough Knowledge of their horizon & tracing by pits from surface might have fully proved the area at comparatively small cost, & before completing Such elaborate arrangements for mining & shipment.

July 31. Start for exploration of Skidegate Straits. Camp beyond the second narrows[222] in a very bad place for wood, which unfortunate as all very wet from almost Continuous rain. Few good Camping places about here the present spring tides rising quite to the Edge of the woods, in fact often covering the lower branches of the trees in sheltered situations. All the dead wood wet & much of it rotten, & owing to the heavy & continuous rains a gentle ooze of water trickling out all along the beaches just about high water mark.

Aug. 1. Continued exploration, turning southward by a channel[223] used by the Indians when *en route* for Gold Harbour, but soon stopped by a dry beach about ¼ m. across, which cannot have more than 4 feet on it at H.W. Turned back & continued out to West Coast by passage to north,[224] which wide & deep, with the swell of the open pacific heaving in at its mouth. Can this be Cartwright Sound of the map? Having estimated distance to outer point, & assured ourselves of the open character of the passage, returned, getting over the second narrows just in time, before the change of tide. Camped in a large bay on the North Side of inlet.[225] Heavy & almost constant showers during day.

The trail marked on Chart as about 3 m. to the West Coast from the Long Arm, comes out to the head of this bay. The valley from this end looks quite low. Trail not much, perhaps never, now used.

Aug. 2. Continued track Survey on opposite side of strait, getting to the E. end of the first or East Narrows about 11 A.m. Found here two Indians who had agreed to come to show us the Coal. This they did, but the deposit not a very promising one, & involved much scrambling about through the wet woods to get at & examine it. Having lunched & satisfied the Indians for their trouble continued survey, getting back by way of channel S. of South Island, to schooner at Fishery about 6.30 P.m. Got dry clothes & at work on notes &c.

Many showers today with strong S.E. winds but not Continuous rain.

Aug. 3. Rowed around this A.m. to Indian village to find an Indian to accompany us to Masset. One named Mills recommended by Smith. Found him at home but the talking appearing interminable asked him to come round to Schooner *tenas polikely*[226] & talk it out. Bought some Indian *ictas* & then Setting sail Crossed the Inlet. Examined junction of aqueous & agglomerate &c. rocks[227] E. of Alliford Bay, & then sailed across to Maude I. Completed track survey of S. side of Maude I & returned on board about 7 P.m. Found Mills waiting, with a number of other Indians, Some wanting to sell Curiosities &c. & others only prompted by Curiosity to roost about the schooner. Had an almost interminable wa-wa with Mr Mills & finally consented to pay him one dollar & a half a day while on the trip, & give him a potlatch of Three dollars in consideration of his trouble in returning by the Shore on foot. Glad to get the decks cleared at last, the cabin fumigated, & quiet restored. One very friendly Siwash with a square peaked Cap, proud of a little English, marched down into the Cabin, lifted a Chart, unfolded it, & began turning his head to one side, pointing places with his finger, & pretending to look very knowing, but really appearing very Monkeyish.

Today almost quite fine, with occasional spells of actual Sunshine & spots of blue sky among the clouds, giving everything a more chereful appearance.

Aug. 4. Remained on board nearly all day, writing up notes, plotting &c. Mr Woodcock[228] arrived from Gold Harbour this Am. having been delayed for nearly a week waiting for a fine time to make the outside stretch between the two inlets. Says he cannot remember a summer with so much broken weather, in this region.

Rocks about Gold Harbour & adjacent inlets according to him nearly all volcanic. Specimens show to be altered volcanic materials felspeltic & dioritic like those seen abundantly further south. In one place in Douglas harbour[229] 'slaty' rocks. The quartz lead worked formerly[230] occurred on a little projecting point, was quite thin, & ran out in all directions to a feather edge. The whole of the visible lead blasted away. The gold obtained by the HB Co. & shipped to England, nearly free from quartz filled three shot kegs (each originally holding 112 lbs of shot) Mr. W. did some prospecting on this old ledge, going down several feet till no trace of vein could be found. Formed favourable opinion of the deposit if large Company to take up & by driving transversely test the <lead the> ground thoroughly. Other veinlets holding gold elsewhere but all *very* small & not at all continuous.

Yellow Cedar A magnificent grove with trees over four feet through & rising up 80 feet <& start a> clear, occurs at the head of a lake above Gold Harbour.[231] Might pretty easily be brought out.

Indian Name of Gold Hr. *Skai-to*[232]

Skatz-sai, or angry waters. Name of Gold Hr. tribe & of the chief also, as usual.[233]

[verso] 2½ point blankets[234] the recognised Currency among the Coast Indians, now equal to about $1.50 Everything is worth so many blankets, even a large blanket Such as a 4 point, is Said to be worth so many *blankets* par excellence. The H.B. Co. & traders even take blankets from the Indians as money, when in good condition & sell them out again as required.

Near the Skidegate village on a piece of flat ground behind a gravelly beach are two flag poles, which were erected last summer. This was done to signify the <function of> conclusion of a perpetual peace between the Skidegates & Gold Harbour Indians, between whom, –

owing to some complicated intermarriage – there had been a dispute as to the ownership of the land, & at various times much blood shed. Flags hoisted on the poles blew away last winter, but the poles themselves 'are there till this day'

Between this place & the village a log set in the ground, about twelve feet high, & carved like the totem posts, marks the grave of a former Tyie Skidegate, who is said to have died very long ago – perhaps forty years – to have been a *delate hias man, Skookumtyie*,[235] & to have had curly hair.

On Bare Island.[236] of the Chart, opposite the Village, the Indians formerly had a fortified or palisaded Camp to which they might retire in time of danger. No trace appears now to remain.

Mr Smith Says he believes now about 250 Indians in all centre at Skidegate Village, though the greater number generally in Victoria. About the same number probably live at the Gold Harbour Village on Maude Island, though this looks much smaller, being quite new, & all the houses occupied. The land on which it stands was purchased by the Gold Harbour Indians from the Skidegates as being in a better place than that formerly occupied on the West Coast. A great number of Indians must have lived about this inlet at one time. Smith says 12,000?? Some estimate of their number at Certain dates Can be formed by ascertaining <for certain do> in how many Canoes the tribe travelled.[237]

Cod fish[238] & *Mackerell*. Saw both taken in cove where fishery established. Skidegate. The cod not very large, but apparently quite the same as in the east. The Mackerell about the same size but a rounder & stouter fish. The Caudal fin not so deeply notched. Colour paler & not so steely & the spotting not so distinct. Evidently a mackerell but a distinct species from the eastern. As far as I can ascertain these fish are now abundant on the West Coast at Gold Harbour, & from them the Indians make a grease which serves them instead of oolachen oil.

Indian villages. Beyond the third narrows, or place which is dry till high water, on the channel followed by the Indians, when going to Gold Harbour, is a large village Called *Chaatl*.[239] Further on, at the mouth of the entrance to Gold Harbour &c., as shown on map, a smaller village called *Kai-shun*[240] Both these villages are now practically abandoned, the Indians living on the E. end of Maude Island as before stated. They are, however, really the former chief places of the So-Called 'Gold Harbour Indians' & are still resided in at certain seasons.

Sea Otters.[241] Are Said to have been abundant on both sides of the Islands, & almost everywhere. None are now got on the inside, the animal here having been completely exterminated. A few are still obtained on the West Coast, especially outside Houston Stewart Channel about the so called *Ninstance illaghie*. It seems however that the Haidas do not themselves often take them, but Chinseyan Canoes engaging in the hunt pay a tribute or toll to the Ninstance Indians for the privilege. So <great> {definite} is the idea of property of certain bands of Indians in certain parts of the country & their right to all the products thereoff.

Succession of Chiefs. A chief dying his next eldest brother succeeds to the rank, or should he have no brothers, his sisters eldest boy.[242] Should neither of these relations exist, the chieftancy drops, & either the consensus of opinion creates another chief, or the most opulent of the ambitious Indians attains the rank by making a bigger potlatch than any of the others. The chief takes a hereditary name on assuming office, becomes *Skidegate, Cumshewa, Skedans* &c. just as ruler of Egypt always Pharo

New potatoes. Well grown, mealy & nearly ripe on Aug. 2. Skidegate. The potatoes planted by the Indians are of late varieties or they might be ready even before this. They are also planted in little irregular patches, the stalks crowded much to thickly together.

Skidegate passage, or strait, all narrow beyond Maude Island, but two Especially narrow places which may be Called the first & second 'narrows' The first of these about three miles long, & averaging not over a quarter of a mile wide. In one place probably not over 200 feet. At high tide appears deep open channel with only a few rocky islets & rocks, but a low water Almost dry for long stretches, with a narrow & crooked channel winding between gravel banks. The second narrows much Shorter, & probably not less than 0.2 of a Mile where least, but very shoal, with several rocks near the channel in the middle. Through both these channels the tide runs with great violence, probably attaining 5 Knots. Tides from W & E meet about the first narrows. A small schooner might be brought through the strait by passing the narrows at slack water, high tide, but probably could not get through both narrows at one tide as the slack water lasts scarcely any time. Our small schooner has passed through, but unless for some particular purpose probably not advisable to use the passage. Another 'Narrows', occurs on the channel turning S toward Gold Harbour[243] & the Indian towns Probably not over 4 feet of water on this at high tide, & dries for a width of at least quarter of a mile. Passage only for Canoes or boats at H.W.

A wide valley runs through to the Second narrows, the water only occupying a gutter in the bottom. Low land, densely wooded thus fringes the strait on each side, slopes gradually up to the foot of

1 George M. Dawson, a formal portrait taken about 1885

2 Geologist Dawson, third from right in the top row, among other officers of the North American Boundary Survey during its 1873 season

3 'Looking east down Houston Stewart Channel from Ellen Island,' 12 June 1878. Dawson had anchored that afternoon in a 'snug bight' on the lee of Ellen Island and then took two photographs, of which this is one. The seated figure, one of those aboard his chartered *Wanderer*, is not identified.

4 'Dolomite Narrows, Burnaby Strait,' 26 June 1878. Dawson found this passage, which he named for its dominant rock formations, to be narrow and crooked, with only six to eight feet of water at low tide. The trees 'seem almost rooted in the beach & their branches hang down thickly nearly to the high-water line.'

5 'Volcanic agglomerates at Point 4, North Shore of Ramsay Island,' 1 July 1878. Uncommented upon in the journal, Dawson's geological interests drew him to these striking formations.

6   'Grassy patch beneath which hotspring rises, Hotspring Island,' 4 July 1878. Haida from Tanu, out collecting birds' eggs, had guided Dawson to the hot springs the day before. He took this photograph of the mossy and grassy patch from which the thermal water flowed, but could not include the pools within the same lens range. The Haida frequently took the waters there.

7   'Clue's village of Tanu, Laskeek Bay,' 9 July 1878. Tanu village possessed about thirty carved posts and sixteen houses. Rain prevented Dawson from taking more than this single photograph of a village which, in 1878, remained in a 'flourishing state.'

8 'Cumshewa Indian village,' 16 July 1878. Dawson was able to take three photographs, two of which are here, emphasizing the poles of Cumshewa. He described the village – with its twelve or fourteen houses, several 'quite ruinous,' and over twenty-five posts – as much more deserted than Skedans and Tanu.

9 'Skedan Indian village,' 18 July 1878. These are two of the five photographs taken by Dawson of the village and poles, which he found particularly interesting. Skedans showed 'signs of having passed its best days' though many of its sixteen houses were still inhabited, and an old woman was putting up a new memorial post for her deceased daughter while Dawson was there.

10 'Skidegate Indian village,' 26 July 1878. Dawson counted twenty-five houses, some in ruins, and fifty-three posts at this large village. His two photographs show as well as any the 'forest' of mortuary and memorial poles which were so striking on the Charlottes during the period of Dawson's tour. The cedar canoes in the foreground are partly covered with blankets to prevent their splitting in the sun.

11 'Basaltic columns, Tow Hill,' 9 August 1878. Tow Hill, near Masset, 'an eminence remarkable in this low country, faces the seas with a cliff composed of columnar volcanic rocks of Tertiary age.'

the Mountains, which then rise steeply. This is also the Case with the arm projecting to the north,[244] on which scarcely any rock exposures occur along the beach. From this a low valley leads through to the Long arm, which is followed by the trail. Beyond the second narrows the Passage takes on the character of the West Coast Inlets generally, steep rocky sides with little or no beach & bold water. The timber at the Same time becomes extremely scrubby on the Mountains, with many dead trees in the woods. Scarcely any soil clothes some of the slopes, among the foliage on which much bare rock Can be seen. The summits are also frequently bare, or show the pale green tint characteristic of bushes &c. as distinguished from the more usual Conifers. These upper slopes look to be grassed from a distance – but are not really so. If originally as thickly covered with soil as the Mountains elsewher they would soon have lost their covering from the slides, which in this preeminently damp country Seem to occur constantly, from water trickling along the surface of the rock. The Yellow cedar begins to abound in small trees after entering the narrow part of the passage.

The axial mountains of the Island, bearing snow still in extensive patches, Cross the passage west of the Slate Chuck Creek & Coal Mine. The peaks of these are not here remarkably rugged. The mountains on the West Coast more rounded & lower, without ruling form. Those E. of the axis apparently composed of beds of the Coal-bearing Series in great part & show long slopes & abrupt escarpments, after the manner of mountains formed of tilted sedimentary rocks. No extensive granitic or gneissic area seen in crossing the islands by this channel. [end verso]

Aug. 5. Weather threatening & showery with low barometer probably indicating wind. Almost decide not to leave, but appearance of Clearing, & light wind induce us to get off not-withstanding the low Barometer. Start about 9 A.m. Go round to the Indian Village & pick up Indian 'Mills' who is to assist in the boat & act as guide. Find a head wind, pretty strong, but continue rowing on against it till after 4 P.m. Boat found to leak very badly, so Camp early, not far past Lawn Hill, to repair her.
Aug. 6. Weather this morning still rough & stormy, an easterly wind Causing a heavy sea. Decide to get off, but on getting outside the little harbour in which we have been, find the water so rough, that with the shoal stony beach it is nearly impossible, or at least very risky to land at points, & bearing Cannot be got out of boat. Run on for a few miles under sail & then rounding a little point, land in the lee, & camp about 9 A.m. About Camp abundance of drift wood, So make large fires & sit

down in shelter from wind & rain. In the afternoon, still no appearance of abatement, pitch tents. Sand in everything, eyes, blankets, food & boots.

Aug. 7. Similar S.E. wind with rain continues, heavy sea piling on the beach, & the water outside covered with white-Caps. Remain in Camp all day, for though might have probably run on safely under sail, could not land, or attend to survey. Got Mills at work giving words for vocabulary, & explaining various Manners & Customs of the Indians.

> [verso] All the Indians, or nearly all about here have a perforation through the septum of the nose.[245] When asking 'Mills' what the use of it he explained by saying *spose hilo connoway Siwash hi-you hi-hi*.[246] This like other marked events in life marked by a potlatch. The nose is perforated at from two to five years old, according to my informant, the father on the occasion *Hi-you mache ictus.*,[247] – or gives away in a potlatch much property. [end verso]

Aug. 8. A fine Morning at last. Off early, & almost immediately get good Sailing breeze, from S.E., without a too heavy sea. Keep on all day under sail, landing only at one place for lunch, bearings, & to examine Clay Cliffs. On landing obliged to haul the boat up & remove all the things from her, Launching her out afterwards through the <small> breakers not without some trouble. Stop at the mouth of a large lagoon, which must be near Cape Fife[248] of the map. Its entrance forms a good harbour for boats, & Can be Entered at high tide. A very strong current, like a rapid river, flows in & out, & at l.w. the channel is Crooked & shoal, to the sea, but good enough for Canoes.

Mills' three dogs, which have followed us till today along the beach – though faring rather scantily – have given up under the quick travelling & long distance of today. When last seen, about noon, were travelling on along the shore, the two larger ones apparently quite understanding the matter & taking the easiest way, the smaller keeping as near to the boat & edge of the sea as possible & looking wistfully toward us. They have probably gone back to Skidegate.

Mills tells me Some strange stories of the superstitious notions of the Indians with regard to Rose Spit, round the Camp fire this Evening. Though much above the average in intelligence he evidently quite believes them.

Aug. 9. Off Early, with very fine Calm weather, & little sea. Row round the much dreaded spit without any difficulty, & find the sea on the west side not heavy. Continued on to mouth of river at Tow Hill,[249] where lunched. Then went about a mile further to sheltered bay where Indians are living making dog-fish oil & drying halibut[250] Mills says

this is the only place to stop this side of Masset, or might have gone further. Got some photos. of the hills &c.[251] Examined rocks at next point[252] where Tertiary Sandstones, with fragments of lignite occur.
Aug. 10. A Magnificent day, clear & warm, & almost perfectly calm. Continue on along shore for some miles, when land to examine rocks. Find fossils abundant stop to Collect them & get observation for lat. Determine index error of instrument. Arrive at Masset about 4 P.m. finding a heavy tide running out of the inlet. Find schooner at anchor. Land & take photo. of Indian Vil.[253] to use up last plate of set. Interview gentleman in charge of H.B. Co post.[254] Pay off Mills & Billy. Engaged all evening putting fossils &c. away, writing up notes &c. Take observation on Polaris for lat, & another for time. To bed at 12.30.

The 'Otter' has been here. Left last Monday for Victoria. My mails are at Fort Simpson, as they did not know where I was, or what plans might be, did not like to bring them here.

[verso] *The Coast between Skidegate & Masset*, in some respects resembles that between Cumshewa & Skidegate. A bare open stretch with no harbour & scarcely even a Creek or protected bay for Canoes or boats, for long distances. The beach is gravelly & sometimes coarsely stony to a point near windbound camp of track Survey.[255] Beyond this it becomes sandy, & though not without gravel continues generally of Sand, all the way to Masset.

Lawn Hill is evidently Caused by the outcrop of volcanic rock described in field book,[256] is probably Tertiary. Beyond this for some distance, & including the region about Cape Ball, cliffs, or low banks of drift-clay, & sands characterize. They are generally wearing away under the action of the waves, & trees & stumps may be seen in various stages of descent to the beach. In some places dense woods of fine upright clear trees, are thus exposed in section, & there must be much fine spruce lumber back from the sea everywhere. Very frequently the timber seen on the immediate verge of the cliffs, & shore is of an inferior quality, rather scrubby & full of knots. The soil is generally very Sandy where shown in the cliffs, or peaty in bottom places where water has Collected. Sand hills or sandy elevations resembling Such, are seen in some places on the cliffs, in section, & there is nothing to show that the Soil away from the Coast is universally sandy, but the fact that the upper deposits of the drift spread very uniformly & are of this character. Further north the shore is almost everywhere bordered by higher or lower sand hills, Covered with rank Coarse grass; beach peas[257] &c. &c. Behind these are woods, generally living though burnt in some places. The trees are of various degrees of excellince, but most generally rather undersized & scrubby. This part of the coast is also characterized by lagoons, & is evidently making, under the <Constant> frequent action of the heavy South East sea.

Rose point & spit is a most remarkable promontory, dependent apparently on no geological feature, but Caused merely by the meeting of the sea from the S & SE, with that from the West, which comes in from the open ocean round the North End of the Island. The <lower on> protunal part of Rose point, near Cape Fife of the Map does not differ in any respect from the low wooded coast to the south, but back from the Shore line are <neither beaches> both lagoons, & lakes, which appear from the Indian account to be very numerous, & neither more nor less than ancient lagoons now filled with fresh water. Further on, the point becoming more exposed & narrower is clothed with stunted small woods, which in turn give place to a bare expanse of rolling sand-hills, covered with rank thin growth of grass. Beyond these, the narrow gravelly point is covered on top with heaps of drifting sand, & great quantities of bleached timber, logs & stumps, piled promiscuously together. Where the point is covered at high tide, this ends, & it then runs on with a slightly flueuous course as a narrow steep sided gravelly ridge. This Slopes away under water, & at the time we passed (tide about 3/4 in) there were two islands of gravel lying off on the same general Course. The sea from two directions rolling together on the shallow water of the spit, with a heavy tidal current running across it, must indeed when wind is added make a very unpleasant Surf for Canoes or boats. Two vessels have been lost on the spit, one a H.B. Vessel[258]

From Rose pt to Masset, the minor indentations of the Shore are so slight, that it may be described as forming one grand Crescentic bay of [blank] miles in diameter.[259] With the exception of one or two small rocky points the beach is smooth & regular, almost altogether sand in Some of the bays, coarse gravel, showing evidence in its steep slope above ordinary H.W. of very heavy sea at some times. Low sand-hills everywhere form a border to the woods, which densely cover the land, & grow thick & scrubby toward the shore, whatever they may do further in. The trees are chiefly *Abies Menziesii*. The water far off the shore is very shoal, especially on approaching Masset, where kelp is found forming great fields far out to sea. [end verso]

Sunday Aug. 11. Writing up notes & attending to various home duties in morning. Went to Church at 11 A.m. summoned by the bell which has just arrived for the mission here, under Mr. Collinson,[260] a fellow worker with Mr Duncan[261] The congregation besides ourselves consisted of two Haidas Several Chimseyans, & Mr Offutt of the H.B. Company. Nearly all the Indians of the place are now away at at house raising & potlatch in Virago Sd.

Dined with Mr Collinson. Got some newspapers in the Afternoon & gleaned many items of interest. Returned to Mr Collinsons for tea, at his invitation, & spent a pleasant evening chatting about things in general. Made preparations for tomorrow, wrote up notes, & turned in.

[verso] *Potlatch.*[262] Mr. Collinson gives me some additional light on this custom.

When a man is about make a potlatch, for any reason, such as raising a house &c. &c. he first, some Months before hand, gives out property, money &c., So much to each man, in proportion to their various ranks & standing. Some time before the potlatch, this is all returned, with interest. Thus a man receiving four dollars, gives back six, & so on. All the property & funds thus collected are then given away at the potlatch. The more times a man potlatches, the more important he becomes in the eyes of his tribe, & the more is owing to him when next some one distributes property & potlatches.

The blankets, ictus &c. are not torn up & destroyed except on certain special occasions. If for instance a contest is to be carried on between two men or three as to who is to be chief, One may tear up ten blankets, scattering the fragments, the others must do the same, or retire, & so on till one has mastered the others. It really amounts to voting in most cases, for in such trial a mans personal property soon becomes exhausted, but there an under-current of supply from his friends who would wish him to be chief, & he in most popular favour is likely to be the chosen one.

At Masset last winter, a young <man> man made Some improper advances to a young woman, whose father hearing of the matter, was very angry, & immediately tore up twenty blankets. This was not merely to give vent to his feelings, for the young man had to follow suite, & in this Case not having the requisite amount of property, the others of his tribe had to subscribe & furnish it, or leave a lasting disgrace in the tribe. Their feelings toward the young man were not naturally, of the Kindest, though they did not turn him out of the tribe as they might have done *after* having atoned for his fault.

Totems[263] are found among the indians here as elsewhere. The chief ones about Masset are the Bear & the Eagle. Those of one totem[264] must marry in the other. [end verso]

Aug. 12. All prepared early this morning to set out for the exploration of the reported great sheets of water above here. Waited for some time for the tide to slacken, as it was running out like a mill race. Got away at slack water, & soon had a good Current with us. Kept distance partly by time, & partly by eye estimate, though difficult to get it exactly owing to the unknown but varying strength of the Current, & numerous eddies. Stopped for lunch at the Mouth of a small river[265] where post pliocene shells were soon found, not far from the waters edge. Deposit in many respects comparable with the Saxicava Sand of the East.[266] Reached the head of the narrow Passage just as the tide began to run out. Kept on a couple of miles, camping in little bay where a nice stream, & remnants of former Indian Camps.[267] Night being Clear, got a good set of observations for latitude & time.

Saw great numbers of wild geese[268] today, on the tide flats, & strips of green grass & weeds which run along the shore just above high water.

Aug. 13. Travelled on, skirting the Eastern Shore of the first great 'lake'[269] Passing the mouth of a large river,[270] which opens in an extensive bay with very wide flats dry at low tide, find several canoes full of Indians looking out for salmon. One of these contained the old 'Doctor' of the Masset Village,[271] with his assistant. The old man distinguished by an immense & dirty mass of grizzled hair rolled up behind his head. This is never Cut, & in it his medicine is supposed to lie. On our approach he shouted 'good day' which soon proved, however, to be all the English he knew. Inquired whether all 'King George men'[272] & on answer being given in affirmative said 'very good' 'Aukook Illaghie King George Illaghie, Aukutty Bostons tike Kapswallow fie mika Klooshnaanich.'[273] He then immediately wound up with the broad hint 'Hilo tobacco?'[274] The old Man further informed us that there were no salmon today, but would be very many tomorrow. Just as he spoke the ripple of one going over the flat was seen, & he & his assistant were off like an arrow in the canoe, which was managed with great dexterity & turned about in an incredibly short space. On overtaking the fish, the assistant hurled a spear, which however missed its mark, falling harmlessly into the water.

Stopped for lunch at the mouth of the Second 'lake'[275] a narrow Passage[276] blocked by a large island,[277] & through which the tide runs with great velocity, especially at ebb. Started round part of shore of upper lake, & camped near the Mouth of the Ma-min River, near an Indian Camp.[278] Went some distance up the river, on which Coal reported to occur, but could find nothing nearer to coal than a few pieces of obsidian. Got thoroughly wet struggling through the woods in heavy rain, & in coming back got involved among small creeks & lagoons of the delta.

[verso] *Tides at Masset. Aug. 12.* Day of full moon. H. water at 30 minutes past noon.
Aug. 13. H. water at 1h 15m. P.m.
   The current runs in up the channel 2½ hours after falling by the shore. Ebb runs about 3 hours after water begins rising on the beach.
   Rise & fall about 14 feet springs, (Est. only)
Aug. 13. Passage to inner, or upper lake.[279] Tide turned to run in at 20 M. past noon.
   The rise & fall of the tide in the first lake 8 to 10 feet in the second or upper, less, probably averaging about 6 feet. [end verso]

Aug. 14. Had made arrangements last night with a brother of our Indian (Jim) to go up the river & show R. the coal, while I went on with running survey of upper lake. Very heavy rain in the early morning. No Indian appeared, & breakfast also late. After some time, the Indian arriving, & assuring me that still possible to go to coal & return before night, decided to carry out old plan. Day continued overcast, with heavy showers, rendering out-door work far from pleasant. Got back to camp about 7 P.m. & found R. & Indians back before me, warming themselves at a fine fire, where we were glad to join them. R. had ascended the river in Canoe some distance to a great log-jam, & then walked Some miles. The 'coal' proves lignite.
Aug. 15. Off early. Crossed the upper 'lake' making some soundings by the way, descended the rapid, – for such the outflow now is, with a strong ebb tide. The water breaks White in several places, & the speed must be nearly ten knots. Continued along S.W. Shore of Main lake, Camping at the head of a long Inlet with an unpronouncable name.[280] Tent pitched on a rather quaking Sod of grass where intersected by numerous little creeks into which the tide flows Got observation for latitude, which much needed here, & then turned in. Day overcast with showers.
   Came unexpectedly today on an old Indian & his wife Camped on a small Island on the lake, the old man engaged in Making a canoe, & a temporary bark house put up near a trickle of water. He actually presented me with a couple of small salmon, in exchange for which complement I gave him a small piece of tobacco all I had. He was very polite, & quite a good example of the better Class of old-time unimproved Indians, different as daylight from dark, from those who have been working on Schooners, or in Victoria, & have learnt various 'white' ways, including the use of oaths & slang.
Aug. 16. Hurried in leaving this morning by the tide, which beginning to run out, threatened to leave our boat dry. Made the round of one more great inlet,[281] & then worked along the north shore till nearly 6

P.m. Camping a few miles west of the main outlet. Day fine on the whole though with occasional showers & S.W. wind. Sky, however, continually overcast, as usual! Our Camp a little north of the Mouth of the River Called *Ain*, on a fine regular beach, with abundance of driftwood at hand. A trail leads over from the beach, about 300 yards to the river, passing through woods in which some fine spruce trees. The stream a large one, navigable by Canoes, though said to be impeded by sticks. Said to flow at no great distance, out of a great fresh water lake,[282] on which the Indians have a canoe, or Canoes.[283] The lake is described as very large, even comparable with this one in size but very full of islands. It is firmly frozen in winter, & is a hunting ground & berry-gathering region. Where the trail strikes the river, are two Indian houses, & a fence & Salmon trap arranged across the river.[284] A neat stage of logs for landing on, little steps Cut out in the bank to the door of the principal house, & other signs of care. No Indians here at present, the salmon fishing not having begun in Most of the rivers.

Aug. 17. Off early, Knowing that must catch as much as possible of the ebb in the long Channel leading down to Masset. Worked along shore to entrance & then down the West shore of Channel, reaching schooner just before 6 P.m., in heavy rain, which has continued most of afternoon, though fine in morning.

A bad headache today rendered work a drag. Evg. wrote up part of notes which have got sadly behind.

[verso] *Aug. 17.* The Indians have some stories of their simplicity when first brought in contact with the whites. 'Jim' Says the first white men they Ever saw Came to North Island,[285] Arriving at the season (Aug. or September.) when the Indians all away at their various rivers Catching Salmon. One Man (?) in the village at North Island saw them & their vessel.[286]

Childish Stories of surprise on seeing various unknown articles, or as one of my Indians put it 'aukutty Siwash dam-fools.' Indian given a biscuit. thinks it wood. encouraged to eat it. no water, finds it very dry. Molasses, tastes with finger, pronounces very bad & tells friends so. Axe, strikes fancy as being So bright, 'like a beautiful salmon skin.' Use unknown. Takes handle out and hangs it round his neck. Gun, similarly misconstrued. Takes flint & hangs it round his neck.

The north shore of Graham Island about Masset, generally low with shoal water extending far off. At Masset {instead of} the wide open bays generally met with find a funnel shaped entrancle, leading to a narrow Passage. Entrance holds some shoal water, & two bars, but navigable passage. Where passage begins to narrow, find three Indian villages, one on E. shore not far in from outer pt. Here principal village at present.[287] H.B. store & mission (the latter now established two years come next November). Anchorage opposite here, but strong tide. The second village,[288] about a mile South of this, on same Shore, the third, on the west Shore.[289] Land all low, no hills, lagoons in places along shore. Generally densely timbered, but reports of 'prairies' here & there in the interior.

The timber does not attain a very great growth along the shore, but no doubt is of good growth where sheltered inland. In proceeding up the Inlet, find, near the entrance to a lagoon which runs back on the E. side to nearly abreast the Indian villages, the land pretty Suddenly attains a greater elevation, forming a flat or gently undulating surface at a height of about 100 feet above the water. This formed of drift deposits, Clays & gravels below, hard bedded sands above. This in many extensive areas becomes broken down to lower country. Inlet slightly tortuous, with average width of 1 mile (?) (say 1½ to ¾ mile) Through the whole length (– miles)[290] the tide runs with a rapid current, especially at ebb, & there is no harbour Capable of sheltering vessels from the tide. Masset or *Maast* island[291] (which has given its name to the region) appears to offer protection, but the water behind it all very shoal (much of it not over 1 fm. at H.W.) & great area dry at l. water. At the point Just inside the island a fair anchorage might perhaps be found, sheltered completely from the sea, but exposed to most of force of current. The 'Otter' said to have anchored here on one occasion.

At its south end, the passage expands suddenly, & a great sheet of inland water is Seen, bordered by continuous low wooded land northward & eastward, by hills, rising to mountains in the distance on the west & South, & studded with islands. The lake lies at the junction of the hilly with the low country. Where bordered by low land the shores are flat, with wide bouldery beaches, bare at low tide, or sandy flats, stretching far out. Where the outline becomes tortuous, & the fiord like projections run up among the mountains, the shores become much bolder, with deep water close to them, & narrow rocky or bouldery beaches dipping steeply away. At the head of the inlets & about the mouths of Some brooks, only are wide flats found. The water in these fiords does not appear, however, to be very deep, differing much in this respect from the inlets of the mainland. on the South side of this 'lake' nearly opposite its entrance, is a narrow passage, with mouth partly blocked by islands, leading into the second or upper lake, the indian name of which means the 'belly of the rapid.'[292] The tide runs through here with great velocity, especially at ebb, when it forms a genuine rapid, with much white water.

Kelp grows abundantly nearly across the channels on both sides of the island, showing that they Cannot be very deep. The second lake is much smaller than the first, & separated from it probably by a rocky barrier. Its Eastern edge is formed as in the former case by low country, while its Southwestern is long & fiord-like, with steep banks. It lies in the same line of junction of hilly & low country as the first lake.

Many streams enter the lakes, as might have been anticipated from the wet character of the country. Of these the largest are probably as follows, & well deserve to be called rivers.

1. Stream flowing into the S.E. corner of the first lake, with large sandy flats about its mouth.[293] This was formerly passable for canoes a long way up, & is reported to head in a large lake,[294] which may lie along the line of junction already indicated. Up this stream the Indians formerly travelled by canoe till they reached the lake, or at least got so far on the river, that they walked across to Skidegate in half a day, reaching that inlet by the valley of the Slate Chuck Creek. A fire has now passed through the forest, & many trees have fallen, blockading the river, so that for a long time the Indians have only gone a short distance up it bear hunting. The same fire has also blocked & Caused to be disused, a trail formerly going from the mouth of the river to the E. Coast near the old indian village at Cape Ball.[295] The Indians say that the fire which Killed the trees at Deadtree point near Skidegate,[296] arose somewhere on the Masset Lake,[297] & this fire it may probably have been which caused the windfall above aluded to.

2. River joining the upper lake at E. end, & coming apparently from the South.[298] On this the Coal is found. Navigable for Small canoes several miles, though much blocked by logs. It probably rises to the S.W., in some of the valleys like those which further north hold fiords.

3. A stream (Ain) joining the lower lake not many miles west of its outlet, on the North shore. This has its Mouth on an extensive gravelly flat, for the most part densely wooded, & fronted by a fine & very regular beach. Several Indian homes are at its mouth, & a couple at about half a mile up it.[299] It is Said to rise in a very large lake flowing out at perhaps three miles from the shore. This {fresh water} lake probably lies between the masset expansion, & Virago Sound & may occupy an analogous position to those just described.

These & many other streams are counted as 'salmon rivers' the hook-bills in autumn running up them all as far as they can.[300]

The exposures of rock on the borders of the lakes are strictly limited to the Western & South-western sides, the opposite shores being composed of drift material. The drift of the Eastern shores is much more mixed in origen than that of the narrower western waters, where the rock fragments are almost altogether local in origen. At Several places a terrace, pretty well marked, & apparently about the level of that formerly measured at Skidegate, seen.

Seems in some aspects not improbable that the 'lakes' along the western borders of the hills mark the former spread of the glaciers from the mountains, at the time the drift deposits, clays, boulder-clays, sands &c. were being deposited. They may thus occupy original hollows, formed partly by Glacial Erosion no doubt but more from the absence of drift materials elsewhere thrown down. No very evident moraines, & must therefore suppose that force of waves sufficient to level any material of this kind as it was deposited.[301]

The rocks appear to be all Tertiary volcanic,[302] in many places & over great areas nearly flat, or gently undulating, & seen in Nearly horizontal beds far up in the mountain slopes at the heads of some of the fiords. In other places the beds are locally remarkably disturbed, as perhaps to be expected in a volcanic region. Much siliceous infiltration & many localities where chalcedony & agates[303] abundant. Considerable proportion of trachytic acidic rocks.[304] A remarkable absence of exposures of ordinary bedded Tertiary deposits, which may be supposed to underlie the igneous, but seen only in the one locality where 'coal' reported, about six miles up the Ma-mit River.[305] Perhaps whenever exposed deeply hollowed away & inlets now occupy the valleys. The hills do not form the boundary Southwestward of the Tertiary area, but are also, as above described, formed of Tertiary volcanic rocks, which, for aught known may extend far in that direction. The centres of volcanic Activity would appear probably to have been among these ranges.[306]

The drift deposits[307] on the whole resemble those of the Coast About Cape Ball &c., being Clayey below, with frequent gravel beds, Sandy above, & All very hard. No true boulder clay recognised unless that of Echinus point may represent. Many glaciated stones on beaches, especially towards heads of fiords. Glaciation found in a few places The *Saxicavas* are probably true drift fossils, of those from the Ma-min R. Some drilling may obtain as to whether not later. Bedding well marked in both Clay & sand deposits, false-bedding common in latter. Differ from drift deposits near Victoria in greater regularity & promenence of bedding of Clays, & comparative absence of large stones & boulders in these.

The Smaller, greenish *Echinus*[308] very abundant in all parts of both lakes, & the inlet below. Salmon the only fish Caught in any numbers. Kelp in many places. The ordinary bladder weed very abundant along shores, also Eel Grass.[309] Probable that oysters would succeed well if introduced, as water probably warmer, & much fresh water, with wide flats. [end verso]

Aug. 18. Sunday. At church, & spent most of day with Mr Collinson. Also visited Mr Offutt of the H.B. Co, obtaining Some facts about furs &c. Took photo of Mission premises, & another of Masset Inlet,[310] in the afternoon.[311]

Names of Indian towns.
Kung village at Virago Sd. (Rough K)[312]
Yā-tzā. New village beyond Virago S. = knife[313]
Ut-te-was. Village near H.B. Co &c. Masset
Ka-yung. Village above last on same side.
Yān. Village on W. side Masset harbour.

Coast between Masset & Virago Sd. Everywhere low, & differs from that east of Masset, in being rocky, or covered with boulders along the shore line. No wide sandy bays occur. The points are chiefly of low dark rocks, which probably all belong to the Tertiary igneous series.[314] The timber seen along the shore, not of great size, but pretty interspersed with open grassy spaces, which often border the sea, but do not run far inland. Water very shoal a long way off shore, & great fields of kelp filling the bays & extending far off the points.

Virago Sound or Ne-din.[315] Wide funnel shaped mouth, contracting soon to narrow passage – only ½ m. wide in one place,[316] & then expanding again to large & magnificent land-locked harbour. Low land, densely wooded, borders the whole inlet, though hills & mountains are not far off to the South & west. Rocks Seen along shore, only near bottom of harbour, & at anchorage cove <opposite> {at} Indian village.[317]

E. Shore flat & long flats With boulders at 1. water. W. shore comparatively bold. Many Small streams, & several of some size. The Nedin River,[318] probably the largest in the islands enters at S.E. angle. Went up it about 2 m. in the boat, its course being moderately straight, & trending a little west of true south. It flows out of a large lake,[319] – probably ten miles or more on longer diameter – at a considerable distance. The lake Can be reached by canoe in about half a day from the mouth, but Indians say that many trees here <filling> fallen across the river last winter. The hollow occupied by the lake Can be clearly seen among the mountains, when the range viewed from a distance.

Aug. 19. Did not get away very early, having to buy several little things at H.B. Company's store & make other arrangements. Followed coast round from Masset to Virago Sound, camping in entrance of latter, opposite the Indian Village.[320] Day nearly Calm, warm, & with only occasional, but very heavy showers. A heavy sea setting in from the west, prevented landing on the part of the Coast setting in that direction. Saw a very large shark,[321] off the outer point. It followed the boat for a little while, & frightened Indian 'Jim', who said that that sort of fish often broke Indian Canoes & ate the Indians.

Aug. 20. Coasted round the greater part of Virago Sound, Camping near its SW extremity. Went about two miles up the Ne-din River, joining at the S.W. extremity.[322] This a large stream, filled with the usual brown water. The day windy, blowing almost a gale from the west outside, the harbour showing many white caps. Entered in the river no wind, warm Sun & apparently different climate altogether. Camped near Some Indians, who shot a bear while we were pitching the tents. Tried to ascertain from the Indians, particulars as to distance of lake &c., with probable wish to see it. Indians speak however of many fallen sticks across the upper part of the stream. Could scarcely get to lake in half a day from the <mouth> mouth of the river.

Aug. 21. Coasted remainder of inlet, & getting round early to the Indian village – now quite deserted for the potlatch at the next (new) town.[323] Schooner appeared in the afternoon & anchored about 5 P.m. in cove opposite Indian Village. At work till late on notes &c.

[verso] *Tow & Tow's brother.* A hill resembling that called Tow on the coast between Masset & Rose Spit, occurring on upper arm of upper lake Called Tow's brother.[324] Story that on some occasion Tow's brother devoured the whole of some dog-fish, which in dispute between them, & that Tow becoming Much angered went away to the open coast where plenty dog-fish, leaving his greedy brother.

Indian food. Indians eat the Cambium layer of *Abies Menziesii*, & the *hemlock* and not that of *pinus Contorta* which also occurs in Some places.[325] Many thickets of Crab apple fringing the shores on the Masset 'lakes' much fruit on them but not yet ripe. Told that next month ripen. Then collected, boiled, allowed to remain covered with water till mid winter when gone over, stalks &c. removed & the whole mixed with oolachen grease *quantum suf.*[326] forming a delicious pabulum according to Indian notions.[327]

Canoes. The Haidas great Canoe-makers. At this season many occupied roughing them out in the woods on the Masset lakes & rivers here & there. Bring them down to villages later on & work away by little & little in winter. They frequently take canoes over from here to Ft Simpson for Sale, getting the coveted oolachen grease, & other things in exchange, together with an old Canoe to return in. Afraid to venture across the stretch <from> {to} Rose Spit, they coast round by S end of Alaska, & run across Southward when fine weather & a westerly wind. Old Edensaw[328] says that formerly when starting for Victoria he frequently had 40 men in a canoe, besides various articles of property.

Number of Indians. Mr Collinson estimates the population of the whole north Coast, Masset, Virago Sound &c. at about 700. No considerable percentage of these Indians Are at Victoria, a few go to Wrangel &c. Villages on outer west Coast, & near North Island now abandoned. *A considerable population of Haidas on S. islands of Alaska.*[329] Not the case however that these the original stock, from which islands peopled. Indians Say that no Haidas here till comparatively lately (Mr Collinson thinks about 150 years ago) Internescine wars then occurring drove Some of the northern bands from the islands to find a home across Dixons passage,[330] & these still entirely distinct in language from any indians of mainland, connected with those of this island, & speaking even the Same dialect. More difference between the Masset & the Skidegate Indians than between them & those still further north.

Fur seals begin to appear in some abundance about the first of April, & the season lasts for about six weeks About 800 skins purchased here each year for the last two years, before that as many as 2600, 2300, 1800 purchased in a single Season. Some years many young fur seals shot in the inlet opposite the village.

Sea otters. Now very scarce. Greatest number purchased in any one season for last ten years about 24, some of these coming from the Alaska side. Shot at all seasons. Hunted as follows. A number of Indians in canoes scattered over the water. Otter seen, Shot at by one Canoe, but not with any intention of hitting, dives. All Canoes concentrate & wait. Otter keeps long under water, but at last comes up. Shot at again, & so on till at last weak & breathless Can keep down very Short time. All canoes now in narrow space, & at last, when the otter can scarcely keep down at all, Some man kills it. He & his fellows in the same Canoe get the whole of the money given for the skin. If any Indian shoots at & wounds the otter, & he is afterwards killed by another, the first has to pay the Second the damage done to the skin by his Clumsy shot, thus very Careful when & how they fire, & no general & dangerous fusilade as might be expected.

Bear Skins 50 to 100 purchased annually. The Haidas not very good hunters. do not kill the bear at the best Season, necessary to follow into woods to do so. In spring attracted to shores by the abundance of succulent young grass &c. Indians then kill many. In fall, when Salmon running, come down on shores to get fish. again many killed.

Forty to fifty Land Otter Skins purchased at Masset, & 100 to 150 *Marten* Skins Annually.

Elk (wapiti)[331] are Certainly known to exist about the north west point of the Island, but very seldom killed, as not followed inland.

Salmon, according to Mr Offutt (squire) of H.B. Co. two runs of salmon. First a small fish with bright red flesh, very good. Begins to run about middle of July & lasts about a month. Not in very great numbers, & not much sought after by Indians as they are then occupied at other things. <About the time> These seem to answer to the 'suckeyes' of the Fraser. About the Middle of August the run of larger 'silver salmon' begins.[332] They are red fleshed & good while yet in salt water, caught about the mouth of the inlet &c., but become hook billed, dog-toothed, lean, & pale-fleshed when up the rivers. They run into all the streams, even very small brooks, are large, & easily Caught, & constitute the great Indian Salmon harvest. The Indians follow them up the inlet. Run lasts till about January. These probably answer to the hook billed salmon of the Fraser, but seem from all accounts to be a much better fish.

Trout.[333] Speckled, good fish, to be found in some streams at all seasons.

Potatoes grow admirably about Masset in the sandy soil.

Barley has been tried experimentally, found to grow to great height & ripen well.

Trees. The *spruce* often attains a good size, & is at times very large. Common everywhere, & well grown away from coast. *Yellow Cedar* not, as far as I can learn, found anywhere in large groves of great trees. Scattered everywhere in hilly district in small numbers. Alder attains fine growth frequently, fringing the shore mixed with the more formal evergreens in many places. Crab apples abundant. Hemlock abundant & well grown. Cedar, fine trees on the flats in some places. [end verso]

Aug. 22. Devoted day to dredging in Virago Sound & harbour, at same time ascertaining depth of the harbour in several places. No very striking specimens procured, & general great paucity of life. Upper part of harbour all mud, with many dead & broken shells, fragments of twigs &c. & but few living specimens. On getting back to schooner found several Indian Canoes about, with various things for sale. Bought some bone pins from one man, also some salmon.

Southerly wind all day, falling to calm at times. Frequent showers of rain, uniformly overcast.

Timber fine spruce timber found a little way back from the Shore, probably every where round this harbour. The Ne-din River probably large enough to allow logs to be floated down from the lake above.

Aug. 23. Leave Virago Sound in the boat, for North Island. A fine morning, though fog banks hanging about.

Sea not very heavy though sufficiently so to prevent landing except in Sheltered situations. See a very large shark, which followed the boat for some distance, occasionally Showing its back fin above water. Length estimated at over 20 feet. Get round to the new Indian Village.[334] shortly after 10 A.m., land, & proceed to make arrangements about getting a new Indian who knows the coast west of this, & dispensing with Jim.

There is quite a collection of Indians at the new town at present, on the occasion of the erection of the first totem post, & potlatch consequent thereon. The Kaigani indians from the S. end of Alaska are daily expected. It is intended to abandon the Village in Virago Sd.[335] as it is found that this place is more in the way of traffic & better suited to the wants of the Indians. Edensaw says Indians from the north are constantly coming over here, but not to Virago Sd. This may be, as this is the furthest northern pt. & is besides marked by a low but conspicuous hill,[336] which may serve as a landmark to canoes making the traverse.

Edensaw, the Virago Sound Chief, & the Masset Chief[337] are both here. The former a decent looking & well dressed old man – though suspected of complicity in the robbing of a Schooner in former years[338] – the latter a stout indian, remarkable from his grey hair & beard. Took photo. of the two chiefs, & of as many of the rest of the people as would come. Most, however, disliked the idea, & especially the women, not one of whom appeared.[339]

Edensaws great village was formerly on the shore S. of North Island.[340] It was abandoned finally about ten years ago for Virago Sd, & now another move is to be made. Long ago Edensaw says his country was at Rose Point,[341] & here his people as he says landed originally when the flood went down.

Got away from the Indian Village at last, & rowed round the long point[342] against a heavy sea & westerly wind. Continued on for some miles, in heavy breaking sea, & then ran into the mouth of a small river,[343] known to my new Indian pilot, & moored the boat in still water. Camped on a snug point, with trees shielding from the wind & plenty drift-wood. Night fine. Got good observations for lat & time, though had to readjust my sextant, the horizon glass having by some means got a little out.

[verso] *First white men*. Edensaw Says the name of the first white man Known to him, or handed down to the present time as having been communicated with by the Indians is *Douglas* Captain of some vessel. This was at North Island. Edensaw thinks, however, that white men Seen before Douglas, & that to the very first, the story below relates.[344] It was near winter when a ship under Sail appeared near North Island. The Indians very much afraid. The chief also very much

afraid, but thinking no doubt that the duty of finding out about the new apparition devolved on him, put on all his dancing clothes, & going out to sea in his canoe danced. The Indians say the men dressed in dark Clothes were supposed to be shags,[345] (which look somewhat human as they sit upon the rocks) & that the unintelligible character & general sound of their talk confirmed the idea. One man would say something & then all would go aloft. Say something else & all would come down. A feat the Indians thought almost like flying. [end verso]

Aug. 24. Tide far out this morning, but managed to get boat down the now shoal river, & out to sea. A fine morning but frequent fog banks drifting past Coast along, landing without difficulty where wish, as now sheltered from force of westerly swell by north Island. Stopped to take photo. of very remarkable pillar like rock, in a bay.[346] Rock quite isolated, over 80 feet high Sloping top covered with bushes. Formed of coarse conglomerate. The Indians have some story about the rock which I cannot exactly understand, but it would seem that according to them it was small formerly, & that some Chimseyan Indians wished to remove it, probably this accounting for the hollow eaten in the base on one side. Since then it has grown very much. The Indian name is Hlā-tad-zo-wōh

Make parry passage & stop at Indian Village on S. shore of North Island[347] for lunch. See on end of Lucy Island of Map the decayed remains of a shark which has been 25′ long. It came ashore dead, or nearly so, & much *grease* was tried out of it by the Indians.

Afternoon coasted up E. shore of North Island to Northern point.[348] A strong westerly or northerly wind blowing but did not experience its force, or that of the waves till we reached the north point, when glad to turn back, & return under sail to the Indian Village, where camped.

Sky clouded during early part of evening, but got a couple of observations on the pole star later on.

Aug. 25. Left tent & most of things at camp. Crossed the passage, examined rocks along the S. shore. Then rowed out to Lucy Island of the map, across a wide bay with a high cliffy Island in it.[349] The swell raised by the prevalent westerly winds & intensified by the Strong breezes of the past few days rolling in about a hundred feet from Crest to crest & breaking very heavily on the beach, where it would be impossible to land in a boat. Got back to Camp against heavy tide about 2 P.m. Spent part of afternoon getting Indian words from 'Harry'. Schooner in sight beating up toward passage. Thought her so close at about 6 P.m. that put things in boat & started out towards her. Got into a tide rip which threatened to carry us out to sea, & wind at same time dying away came to conclusion that Schooner would not get in this evening, & might be so far off as to interfere with tomorrows work if we were on board. Put back & slept at old Camp.

Rain in the night

[verso] Seen from a distance, the hill *Tow* does not look so abrupt as when Seen from near the shore on either side. It appears to be separated by a small gap from a second low broad hill, which is probably that seen South of Rose Spit, on the E. shore, but here appears in line with Tow on the spit, or nearly so. This appearance has Caused these hills to be drawn as they are on the map.

Coast between Virago Sd & North Island. Generally low land, with occasional rocky cliffs of no great height along shore, but generally alternating low broken rocks & gravelly beaches. Few Sandy beaches occur. Some rocks seen at a little distance off shore, but no sign of wide shoal belt like that of E. of Masset. Trees along Shore of the usual Character, generally scrubby, owing to their exposed situation. Green grassy patches along the edges of the forest, & on sandy & gravelly spits of old formation. The remarkable rock pillar occuring in one bay has been described elsewhere.[350]

North Island is all composed of low land, probably in no place rising over 300 feet. The country to the south Similar, though higher hills appear about abreast of Frederic Island on the West Coast. [end verso]

Aug. 26. Found the schooner at anchor in Bruin Bay. Put things on board, & then set out for the abandoned Indian village[351] from which trail to the west coast starts. Took Harry as guide & Crossed by the trail, which little more than ¾ mile to bay S. of Cape Knox.[352] The trail a very devious one, over & under moss-grown logs & through thickets, all soaking wet from the frequent showers. Come out at the mouth of a little creek on a fine sandy beach on which the waves are breaking with a regular steady roll, from end to end. Went along shore to next point south,[353] from which got bearings & sketch of Coast as far as in sight. The Shore remarkably rugged, here & southward. Broken rocky cliffs & rocky pinnacled islands, with reefs still further out. On all these a great ocean swell forever breaking & the water yeasty with foam, & a mist hovering along the whole Coast-line from the spray of the breakers. Ate our lunch in a shower of rain at the mouth of the Creek, & then returned by the trail. Found no boat, & after waiting some time set out along the shore toward the bay in which schooner lies. Found the boat high & dry on the beach, having been Carelessly allowed to become so by the men cutting wood. Got put on board the schooner by an Indian

Canoe which fortunately there. Boat brought back by men after 9 P.m. when the tide rose again.

An unpleasant anchorage as so open that the swell from the Eastward & Caused by the tide rip keeps the Schooner in a state of perpetual roll.

Aug. 27. By time water casks had been filled, no wind in harbour waited some time with sails up, then got off with light air, but carried away by the eddy. Had to get boat out to keep schooner off rocks. Getting out of eddy got into main tide, which breeze not sufficient to enable us to stem. Made the round of Lucy I. & after an ineffectual attempt to get out against the tide in the North Pasage,[354] Came to anchor opposite old Camp, & new Indian Village. Got off at slack water, after a few hours, though again nearly carried down on rocks by tide. Proceeded Eastward with light wind & strong tidal current. Got a couple of soundings in the evening.

Aug. 28. About opposite Virago Sd. this morning & continued on general north eastward course, with light easterly winds all day. Got two Casts of lead at 111 & 130 fathoms in places pretty well fixed by bearings, though weather thick. Also had the dredge down for about half an hour at the former locality, & brought it up full of sand & brittle stars, with a few shells. Attended to some duties about outfit. did a little plotting &c. one very beautiful spiny star fish. Carmine-coloured[355]

Evening nearly becalmed in gently rolling swell.

Thermometer thrust into mud of dredge went down to 47°

Aug. 29. In sight of Zayas Island early this morning. Under Sail all day, getting inside Port Simpson about 8 P.m. Light westerly winds & fair weather with occasional glimpses of the sun. Views of the Southern promontory & islands of Alaska through mist & clouds which seem constantly to enwrap them. Passed pretty close to Zayas Island, to the north, also to N. Shore of Dundas Island. These outer Islands, as before remarked further south, generally low, & seem to show slope of surface of older cryst. rocks westward. Landed on an island of the 'Gnarled group'[356] & got specimen of the rock, also little vein with some copper pyrites, & evidence of glaciation Seaward.

Rocks of all these islands appear to be granitic or at least wholley-crystalline.

A very fine view looking up Portland Inlet as we Came up this evening the mountains on its E. side Singularly bold & abrupt. A fine peaked Snow-Clad Mountain[357] seen behind Port Simpson, the same on which a bearing formerly obtained from near Rose Spit. Like the rest shrouded in grey clouds & seen only partially & occasionally.

The rock of which note made on the Chart this morning, off Zayas Island, is probably the 'Devils ridge'[358] of which the position is doubtfully indicated on the chart. At the time Seen, breaking heavily occasionally, but not often. The sea moderate, & the tide nearly full. It would be a dangerous rock for vessels entering the channel for Port Simpson, from Dixon's Strait.[359] Nor is it alone, for with the generally low character of Zayas & other islands, the bottom does not appear to deepen <good> fast or regularly. Breakers observed far out from Zayas Island in several directions & little rocky islets near shore.

Got large mail here, the first we have reached since leaving Victoria. Up late enjoying letters & looking over papers.

Aug. 30. Spent most of day with Mr McKay of the H.B Company looking at his specimens & gathering items of interest about the Coast. Got a photo. of the village of Port Simpson, R. afterwards getting another of the harbour from a hill[360] Visited Mr McKay's 'Chimseyan Lode'[361] The H.B post or fort[362] here much in the usual style, in good repair. Building painted white & red facing on a quadrangle, Surrounded by a pallisade which once had bastions. One of these still standing <last> used as a hay barn. A loft of Chilcat Indians from the Country bordering on the Esqumaux, in Alaska arrived to trade today & all day examining goods & haggling over them. Brought down a valuable lot of furs which they have themselves bought from the interior indians.

G.M. DAWSON TO MARGARET DAWSON, 30 AUGUST 1878, PORT SIMPSON

by the superscription you will see that we have left behind us the North American Hebrides, & have again got to a place with a name, the northern point of Brit. Columbia on the Coast.

We arrived here yesterday evening after a pleasant passage from North Island. It was dark when we got in but R. went ashore to hunt up our mail, & we had the pleasure of overhauling the first home news we have had for three months back. The latest Montreal, – or rather Metis – dates are Early in July, as there has been no steamer here for some time. I have had bad luck in Sending letters, & find a note written from Masset, to William, still lying at the H.B. Fort here, no chance having occurred of forwarding it. Thus it is that being quite uncertain how long this may lie here after we leave, that I do not propose now to do more than report our welfare.

We hope to get away tomorrow on our way southward, looking at a few places *en route* & stopping perhaps some time near the north end of Vancouver Island if the weather holds good. Rain we are now quite accustomed to & everything has a dampness about it, so much so that if I put away a pair of boots for a few days, on looking them up again they generally have a coat of mould outside & in.

The Chances are so very infrequent of communication with Victoria from here, that I fear Rankine may either have to leave very soon, or be too late for his work. I hope to know more about this tomorrow & shall add a note should anything occur. There is quite a large village here, inhabited by Indians under the guidance of a Mr Crosby,[363] a rather imprudent Methodist Missionary. He has a large white church on the slope above the town, which is quite a landmark from far off the harbour.

Please give my best congratulations to William on his good final Standing.

Aug. 31. Had intended leaving today, but a strong south-easterly gale, with squalls & rain in progress, & judged it better to remain. Went ashore & made arrangements about some additional supplies – Corned beef & butter – Afternoon went round harbour in boat examining rocks, though the strong squally wind, with rain, rendered it very unpleasant, & quite a disagreeable sea had got up toward the north, or outer end. Shot a few small plover[364] on the way back. Wrote up some notes &c. Evg. paid a visit to Mr Mckay of the H.B. Co, getting on board rather late, & sitting up reading till nearly twelve.

Sept. 1. Weather appearing more moderate, got away this morning, but found a very strong head wind blowing Beat southward slowly, arriving after dark opposite Metla Katla. Weather thick & afraid to run in. Beating about outside all night, with lighter wind.

Sept. 2. Found ourselves early this morning back, about abreast of Ft Simpson. Got a more favourable wind, & reached Metla-katla about 10 A.m. Visited Mr Duncan, who received us Cordially. Afternoon went round the harbour, examining the rocks, & got photo.[365] of Part of the town. Evg. spent with messrs. Duncan & Tomlinson.[366] R. decides to stay here to catch first steamer for which Mr Tomlinson also waiting. Up late getting things packed & arranged.

Sept. 3. Intended to leave this morning but weather turning bad, not withstanding very high barometer, decided to wait. Weather continued to grow worse, gale of wind with very heavy rain showers from the South East. Mr Tomlinson Came on board, & remained talking till about 4 P.m. Had a visit from a Capt. Madden[367] & a man Called Jones, the latter claiming to be an old acquaintance, having been on the Boundary Commission, in Ashe's Party.[368] Told that the 'Grappler' stopped at Inverness (Woodcock's landing)[369] early this morning, & promised to be back on her way to Victoria in three days. Took tea at the mission & spent part of evg. there, leaving R. in hope that we may have early start in morning.

Engaged an Indian boy this morning for extra help in boat &c. Wages to be $20.00 per month, not less than one month paid, & $10. return fare from Victoria in steamer.

Sept. 4. Morning promising fine weather, & South Easterly Gale having ceased, decide to get away. Land to see about barrel of oolachens which I had agreed to purchase, & find R, with Messrs Duncan & Tomlinson at breakfast. Get the fish, & shortly afterwards get away, though after leaving the harbour almost constantly in sight of the Village till night fall, as we beat against wind & tide, making very little. The current appears at present to set out Chatham Str.[370] northward at both flow & ebb. This may be owing to the Volume of water discharged by the Skeena, or to the present prevalent South-Easterly winds.

Had the dredge over opposite Metla Katla in about 50 fathoms, & brought it up full of slimy mud, & stones, some several inches in diameter, mostly more or less water rounded, but with occasional angular fragments. In the mud found one living & several dead *Rhynchinellas*, a few *Ledas*, a brittle star,[371] & one or two other shells. Life very scantily represented. Such a deposit as this might almost form a 'boulder clay'. Have not before struck similar bottom, & judge it not improbable that the stones may be brought down by floe ice from the Skeena in spring.

Wrote up notes, did some plotting & attended to Several other little matters which have got behind.

Sept. 5. Found ourselves this morning off the Skeena, with Inverness in full view, though at a considerable distance. Light baffling winds all day, with heavy tide running out against us most of time. Getting abreast of White-Cliff Island,[372] where some men are trying to open a marble quarry, went ashore for a few minutes to inspect it. Toward evening got a little puff of fair wind, with rain, & distant thunder. This pushed us on nearly to the south end of Kennedy Island, when again becalmed. Did a little plotting, reading &c.

Sept. 6. Drifted back last night with the tide, which appears to flow in by Ogden Channel at the flood, got into Chalmers anchorage, abreast North End of Kennedy Island. Morning Calm. Went ashore for water, & got additional supply of wood. Light breeze springing up from North or North west between 10 & 11 Am, got away, & have had moderate to light fair wind light astern ever since, making a fair afternoon's run. Did some plotting, wrote up notes, read, & attempted to colour my Sketch of Portland Channel.[373] A very fine day, with much sunshine, & the moon & Jupiter now shining brightly ahead.

> [verso] *The Chimseyan Indians* are closely related to the Tinne, & have in fact come down from the interior onto the Coast by the Skeena River. The Skeena is not the real Indian name of the river, which is differently pronounced, & the name Chimseyan means simply peo-

ple from the Skeena.[374] Mr Hall[375] here at the H.B. post, who knows the Carrier language well finds many Collateral or similar words between it & the Chimseyan. The migration did not take place within the traditional memory of any Indians now living, but may not have occurred more than about 100 years ago. The Chimseyans displaced the Tongas Indians, who now occupy the Coast from the W. side of Portland Inlet to the Stickeen.[376] Their country being Part of that of the Kaigani, or migrated Haidah Indians. The Haidas have always been in the habit of resorting to the Nasse to fish the oolachen, the Chinseyans allowing them to do so, or rather fearing, or being unable to prevent them.

Haidas. Mr Hall of H.B. Co tells me that a Custom among them that when a girl arrives at puberty, she goes about for a time in a peculiar Cedar bark Cloak, which conceals the face. Afterwards a feast or time of rejoicing occurs[377]

Pholas borings.[378] *North I*. In the large bay next north of Parry passage on e. shore, on S. shore of bay. Found distinct pholas borings in Calcareous shales, above present H.W. mark, & {altogether} above the position in which these shells would now live.

Oolachans on the Nasse. The first & great run occurs about the Middle of March, a second smaller run is said to occur in June.

Coal. An indistinct report of Coal on Work's Canal,[379] but appears unlikely to be true.

Distribution of Cervidæ.[380]

Moose.[381] Are found to within about ten miles north of Ft St. James, Stuart Lake, down to Ft. George, & in the whole country North East of the chain of lakes of which Stuart L. is one.

Cariboo[382] very abundant E. of Ft George, also on hills West of the Quesnel & Blackwater trail. Extend to the 49th parallel near Okanagan in winter. Abound about the head & N. side of Francois Lake.

The small red deer,[383] formerly <only> found northward only to Ft George & not common there, now abundant about Ft Fraser, & found also on Stuart Lake near the fort (Ft. St. James).

Grizzly bear,[384] said to be two distinct kinds, large & small, but with uncertainty about specific lines among bears this not of much value.

Copper Shown fine specimens of Copper pyrites & bornite[385] by an Indian. Said to come from a place a little below the Forks of the Skeena, & to exist, of course, in great quantity. Indian believes the locality unknown to whites.

Gold Capt Madden has a few small quartz specimens with large pieces of gold scattered through them. He says the ledge is well defined, & he hopes to be able to do something with it. Locality about 60 m. up the Skeena on the left, or south bank. The gold 'free' & little pyrites visible.[386]

The several blocks of land separating the mouths of the Skeena {& adjacent channels} with the exception (only?) of Digby Island, are generally high & mountainous, rising steeply, or with a very narrow sloping foot from the water's edge. Kennedy Id. is also of this character, exceedingly bold land, sloping down into Arthur passage at a very steep angle. On this face two great bare 'slides', showing solid rock from base to summit {(showing granite apparently)}, one of them connected with a brook. These slides precisely like those seen frequently in the Queen Charlotte Islands.

Grenville Channel opens rather widely at first, but gradually Contracts, the hills immediately bounding it, are at first not very high, are densely wooded, further back, however, both on the Mainland & Pitt Island Mountains still holding extensive patches of last winter's snow, in deep drifts, occur. The summits of these are bare & treeless, & even where they show no snow, whitish, from the almost continuous exposures of granitic rocks. The views up some of the valleys & glens which lead streams from there down into the Channel, singularly wild & beautiful. [end verso]

Sept. 7. Nearly out of South end of Grenville Channel this morning at breakfast time, having been favoured during the night by breeze & tide. A light head wind springing up, however, set us to beating, & remained beating about mouth of Channel most of day. Even when tide running out strong below, a surface current, probably impelled by the wind, & only a few yards deep, kept dragging us back into the strait. Landed, & got a photo. of Channel from Yolk Point.[387] Anchored for a time in a cove, as we continued to drift back, but finally in the afternoon, getting a fair wind, set off, & now, (8 Pm) have Crossed Wright Sound & are fairly in <Tolmie Channel> McKay Reach

The 'California' passed us close, on the way up today. Still no news of the 'Grappler' or 'Otter'.

[verso] *Most of the land about Wright Sound*, rises at once steeply, & without beach, from the water, to mountains of considerable altitude. Some of these, especially that on Gill Island, & on south end of <Grenfell> {Hawksbury} Island, are peculiarly picturesque in form. The latter range, <called the Wimbledon Mts. on the Chart,> slopes to the North west, breaking off in a series of abrupt step like escarpments in the other direction, almost as though dependent <on> for form, on bedding of massive strata.

These mountains of the off lying Islands, are however pigmies to the serried & snow burdened ranges which form the axial Summits of the Coast Range. Looking up Douglas Arm,[388] from Wright Sound, tier upon tier of these, glowing in the pale rosy light of a rather cold sunset, appear. Fields of driven snow of great size, & evidently in some places of immense depth, shroud their Summits, while here & there a long ridge, or sharp Crag stands above the white Surface. It would almost appear that some new snow has already fallen this autumn on these unnamed & unmeasured giants. [end verso]

Sept. 8. Beating about, with light baffling winds, in McKay's Reach all night, & a considerable portion of this morning. Could see ahead of us all the time a fine breeze drawing into & down Fraser Reach. Finally, by aid of the sweeps, got into the wind, & have ever since been making good progress Southward, being now, 8 Pm, not far from entrance of Tolmie Channel. Did a little sketching & some reading today though generally lazily inclined. Weather remarkably fine, & the scenery wild & magnificent.

'Grappler' passed us on the way down at 9 P.m.

[verso] *McKay Reach, Fraser Reach & Graham Reach.*

The first named, wide, & transverse to the general course of the inland waters about here. Some fine mountains, of which the peaks still hold a little snow, on S. side (see sketch)[389] Mr McKay Enformed me that schistose rocks occurred here, but those I have seen from the schooner for the most part evidently granitic, gneissic, or dioritic = massive & white. Fraser & Graham reaches, are really parts of one long narrow passage, which in its character is more rugged than any of those we have yet been in, in coming from the north. The mountains Surrounding the channel, are not very high, but some still hold masses of snow on shady exposures. With the exception of the less height of the mountains, this channel resembles the upper reaches of some of the fiords. Many of the Mountains are almost bare massive blocks of grey granite. Where they are covered with timber it is small. (except occasionally along the water's edge) & scrubby. Cascades fall in on every side, & the roar of a large waterfall fills the quiet mountain fenced passage opposite Work Island.[390] This waterfall is the overflow of a lake,[391] which by the apparent size of the hollow in the mountains must be large. A little further on a second stream[392] of considerable size flows tumultuously out, (also on the W. side) & here a party of Indians were Camped salmon fishing. They Called to us, but did not come off.

Rocks almost everywhere in higher Mts. appear massive granitic. Along shore noted strat. rocks, gneisses or mica schists, in several places, but did not think it worth while to stop to look at them, from their monotonous character. On North West side of Work Island, rocks apparently schistose, & seem to include a bed of brown-weathering limestone.

Glaciation noted as rock rounding & distinct & heavy grooving in many places, in the channels traversed today. This both at the water's edge, & a thousand feet or more up the mountain sides. Direction, as might be expected always parallel to Channel, though frequently grooves slope up, or down, <as regards it> on nearly vertical Surfaces.

The point between McKay Reach & Fraser Reach, is heavily glaciated as though by ice coming out of passage to North. [end verso]

Sept. 9. Floating slowly along without wind in Klemtoo Passage this morning. Sun bright & perfectly Calm. Went ashore on a small island & took a photo.[393] looking north up the passage; which if it developes should show curious effect of kelps & reflexion of trees in water. Seems, however, rather an act of faith to expose an 'Extra sensitive' plate ten or fifteen seconds & expect to carry away a picture! Getting a little wind, at length beat out, & now floating about, again becalmed, in the Centre of Milbank Sound. Ocean all open to the S.W. but scarcely any swell.

A lovely night, full moon & many stars. As last rosy light faded from sky, the clamorous din of sea-fowl on the water, & distant rocks, with the occasional sharp snort of a whale blowing at the surface, or the distant sound of one breaching – combined to give a peculiarly wierd effect. <to the> A distant unvisited haunt of the creatures who congregate where man is not. Sketched out plan of report. Reading &c.

Sept. 10. Nearly becalmed all night, & calm continued with Scarcely a break till noon. Bright sun & smooth water. Drifted a little this way with the tide & that way with the wind, but made no real progress. After noon a nice little wind rising, sailed into Seaforth Channel, & Just about sundown anchored at the Kil-Kite Indian Village of the chart, behind Grief Island. The Indian name of the village is Kā-pa,[394] according to Charley *Ham-Chit*, (the latter being the Indian name. This man is the chief, & Came off to us on our arrival with a neat wreath of red-stained Cedar bark about his head. He seems very intelligent & had a long conversation with me after supper.

Sent a couple of young Indians off to Bella Bella with a note to H.B. Store asking them to keep any letters which may come up in the 'Otter'. Promised them a pound of tobacco for their trouble.

Got the boat on the beach to try to find out where she leaked, but could not succeed in discovering any large break. She begins to be rather

frail & strained throught, by much contact with rocks & hard usage.
Sept. 11. Made an early start in the boat, with the Indian above mentioned, & his wife, in a small canoe. The Indian to act as guide to Mr McKays 'Hebrew' mine. Rowed up Ellerslie Channel[395] of the chart a long way getting to the mine about ten o'clock. Examined the little tunnel which has been driven, the ore at its mouth &c. Had lunch, & at noon, set out on return. Strong head wind, Causing us to have a long & heavy row back, getting to the schooner about 6 P.m. A fine day, with a good sailing wind in the afternoon, had we been travelling. Found the Indians who had carried my letter to Bella Bella last night waiting to be paid, also an Indian wishing to Sell a deer – for which he got a dollar – another with some halibut &c. Our guide last night asked me for some old Illustrated news. Today he wants a Cup of flour to make paste to stick them up in his house! Probably Should we wait here tomorrow he will be along asking a brush.
Sept. 12. No wind in the early morning, & very light air for some time after it began to come in. Beat about the entrance to Ellerslie Channel, but did not get fairly out of it till noon. Then got a good breeze, which pushed us up Seaforth Channel, bringing us to Bella Bella about 5 P.m. Went ashore in boat to leave a few letters, & found a note from R, with a few papers of late dates he had picked up. Looked at some specimens of ore Mr Leighton[396] had, & then set out after the schooner. Light variable winds during early night. A very fine day, quite summer like

G.M. DAWSON TO MARGARET DAWSON,
12 SEPTEMBER 1878, BELLA BELLA

Here we are in the schooner, again at Bella Bella, but this time on the way southward. I have no certainty when this note may go, but as some steamer may pass, will leave it, together with another, written long ago but never sent.

I left Rankine at Metla Katla, well provided I believe with money to carry him back to Montreal. The 'Grappler' on which he was to take passage, passed us in the night some time ago, & is now no doubt in Victoria. R. will miss the steamer of Sept 10, but should get home about October 1st. I requested him not to get off the train & spring on again at the last moment more frequently than absolutely necessary to preserve his mental & bodily health; but noticing that he received all these hints with considerable reserve, believe that he made up his mind to do just as he liked.

The weather is magnificent now, in fact the finest we have had yet. Warm, clear & without any fog. The only drawback is the lightness of the winds, which renders sailing rather slow.

I have not yet quite decided exactly how long I may stay at various places on the way to Victoria, but as the season is already so far advanced, judge it best to spend what fine weather remains on the way. There is no time for any separate expedition after return, & I am now provided with everything necessary for work along the Coast.

I do not intend to be very late in returning to Victoria, even in the case of continued fine weather, as I have several things to attend to there which may take some time, & must besides be in Montreal before very long to see about revision of report written last winter, which was left in an incomplete state.[397]

[verso] *Bella Bella names of stars &c.* Stars *To-toa* Moon *No-si.* Sun Klik-si-walla orion's belt, Il-i-wha. Pleides Il-i-wha-so Great Bear Klak-tsoo-wis. North Star Pāice (They are unacquainted with the fact that the north star does not move) Kwa-Kum probably the Dog star. (Sirius) Milky Way Kum-e-e

When asked as to *origin of Indians* Hain-chit gave me the following. Very long ago there was a great flood, the sea rose above everything, with the exception of three mountains.[398] Two of these are very high, one near Bella Bella, & the other apparently N.E. of it. The third is a low but prominent hill on Don Island Called by the Indians Ko-kwus. This they say rose so as to remain above the water. Nearly all the Indians floated away on logs & trees to various places, thus the Kit-Katla's floated to Fort Rupert, & the Fort Rupert's to Kit-Katla. Some Indians however appear to have had small canoes, though the making of canoes not very well understood in these days. These anchoring their canoes, came down when the water Subsided near home (The story a little confused here & the precise use of the Mountains not evident) At any rate, ther remained at last of Bella Bellas just three, two men, a young woman, & a dog. One of the men came down at the village where we are now anchored, another at a village site near Bella Bella & the young woman & dog at Bella Bella. The young woman slept & the dog married her, giving rise to a being half man half dog. Similar creatures soon multiplied, & these eventually turned into the Bella Bella Indians.[399]

Fire, first given to the Indians by the deer. This animal it would appear showed them how to use the fire drill.

When the flood went down, there was no fresh water, & the Indians did not know what to do. The crow however showed them how, after eating to chew fragments of Cedar wood, when water came into the mouth. He also it would seem, by & bye showed them where to get a little water by digging, & soon a great rain Came on, very heavy & very long, which filled all the lakes & water Courses, so that they have never been dry since. The water however is still in some way connected with the Cedar, & the Indians Say if no cedar no water in the country. The reverse at least would certainly hold good.

Ham-chit says the Indians are always talking among themselves about their decrease in number.[400] Long ago he Says they were like the trees, in great numbers everywhere. The fought among themselves (as he said the white men fight among themselves) Some were killed, but always more were born & the whole country teemed with them. Now he says the white men have come & the Indians *Chaco mamaloose*,[401] *Chaco mamaloos*, & Soon there will be none. He pointed out to me the former extent of his village & contrasted its present shrunken Size. Yet he says there is plenty food, plenty fish, & we have various things from the white men which we did not know before. The Indians do not fight among themselves or with the whites, only for a few years was whiskey introduced among them, not long enough to do much harm & yet they die. The Indians he says do not know how to explain it, but as he says – *Klunas saghalie tyee Mamook*.[402] [end verso]

Sept. 13. Got a good fair wind after getting through Lama passage, sailed down Fitzhugh Sound, finding ourselves in the morning not far north of Safety Cove. Wind falling & then coming ahead, kept us beating most of day. Well outside the point by Sundown, & shortly afterwards a good breeze coming up, Made fine progress for a couple of hours, when again becalmed. A magnificent day. Saw at a distance six large Canoes of Newitti & Ukultaw Indians[403] on their way to Bella Coola to trade Blankets for potatoes. Wrote first part of notes on Haidas.[404]

Sept. 14. Becalmed, & rolling about in the open <between> off Cape Caution, most of day. Light air in the afternoon enabled us to make the land, but strong tide running out obliged us to anchor in Shadwell Passage

A fine & warm day, enjoyable enough but for the Sound of the booms & sails lashing about as the rollers passed under us. Saw several sea lions & a couple of large whales[405] playing about. Boarded this evg. by some Nawitti Indians, made enquiries about coal reported by the Chief *Chip*, who is now off with the Bella Coola party. A magnificent night, bright moon & stars, nearly calm.

Sept. 15. Interviewed a couple of Indians about the coal reported near Cape Commerell,[406] but as the story indefinite & distance great, – at least twelve miles – & the locality on a very open coast where it might be difficult to effect a landing, decided not to visit it. Waited till about 11 A.m. for tide to turn, got some wood & water on board, & eventually got away, beating up the strait against a westerly wind. Now at 8 P.m. Rolling about a little beyond Cape Commerell & some miles offshore. The wind has brought us so far but left us in *the lurch*.

We moved yesterday evening on the turning of the tide, from our exposed anchorage in Goletas Channel to a snug cove at the Nawitti Indian Village. For several hours this morning the most doleful crying & wailing was kept up by some women in one of the houses. Learned that this the ceremonial mourning for a little child who died a few days ago, & has been some days buried – or housed – The women were relatives & went on with their work <while> more or less steadily while uttering these heart-rendering cries. A sort of wailing, mingled with interjections, & sentences probably referring to the deceased.

> [verso] *Nawitti Indian Village.* The houses not built on the old ponderous style, whether that may ever have been in vogue before or not. [*Illus.*] No carving indulged in though rude painting of the flat fronts of the houses has been practised. Now very dim. The best marked design represents two 'heraldic' birds, in black & red fighting, one on each side of the door. One small carved 'totem post' in front of the Chief's (*Cheap*) house. Also one other pole, with a cross stick & two men upon it, intended apparently to represent a mast & yard with two sailors aloft.[407] [end verso]

Sept. 16. Rolling about all day, drifting a little one way & a little the other with the tide, but not enough wind to fill the sails. Dull gloomy weather. Begin to repent that I ever set out on this Quatsino expedition, which seems about to involve such loss of time, discomfort, & possibly bad weather. Towards evening a good breeze springing up got away again, but no certainty of its continuing. Decide if not much further advanced tomorrow morning to give up the Quatsino expedition.

Had the dredge over twice, in about 15 fathoms, but got little of interest. The water shelves away very gradually here, & the rapid tidal Currents appear to keep it quite clean. Composed of gravel & rounded & smoothe stones, with little bryozoon or incrustation. A few shells, mostly dead – & many small brown holithurians, sea eggs, & star fish[408] of common beach varieties.

[From 17 until 28 September, Dawson and the *Wanderer* explored Quatsino Sound on the West Coast of Vancouver Island. This portion of the voyage, being off the Inside Passage, is omitted here. It can be found in full in Dawson, *Journals*, 529-35.]

Sept. 28. Rounded Cape Scott at 4.30, & hauled up along the North Shore, for Goletas Channel. Barometer going down & wind continually freshening, till eventually blowing half a gale, & obliged to take in the fore sail. Weather dark & thick & all appearance of a regular Southeaster, which, had it come strong enough might have blown us offshore altogether. Aided by a strong tide however beat in a few tacks into the Channel, coming through a heavy tide-rip on the bar. Made out entrance to Bull Hr, ran in & anchored at Midnight. Found another schooner, the Kate, here, going to Skeena with coals.

Sept. 29. Blowing a gale all night & morning, but Calming down considerably, with signs of Clearing weather, & a rising barometer, about noon. Wind dead ahead for us, however, & the Ebb tide running all the afternoon so that useless to go out. Working at map, reading & writing. P.m. went in boat to examine rocks about harbour, & also crossed the narrow neck of land at head of Hr – about 300 feet – to Roller bay on the exposed outer coast. The crossing a difficult one as the underbrush consisted of a tangled & extremely dense growth of sal-lal, crab-apple &c. from six to twelve feet high. On gaining the coast however well rewarded for trouble in finding a magnificent sea falling in against the upper, or steeply shelving portion of a shingle beach. The simultaneous advance, rise, & tumultuous break of the great blue seas, as they arched up, fringed with little rainbows as their edges became fretted & misty, truly grand. The impressive Sound of the Stones & pebbles along the whole beach roaring as the broken wave retired brought vividly before one the process of the destruction of continents, & the immense sum of work which must be performed by an agent like this eternally busy. The scene almost realized that of a dream of great waves breaking on a beach, which I remember once to have had. No explanation can be framed of the sentiments Called up by the display of such never ceasing force, Can only fall back on Tennyson's – Break Break Break on thy Cold grey stones oh sea, & I would that my heart could utter the thoughts which arise in me.[409]

The distant roar of the Surf on this island appears to Surround us on every side as we lie in this little land-locked harbour in the still night.

Sept. 30. Ran out this morning intending to beat eastward with the flood tide, but, though Calm enough in harbour, found a fierce South easter blowing outside. Obliged at once to take in foresail, & as no prospect of making anything beating against such a wind, ran back into Harbour. Reading & writing most of day.

Oct. 1. Blew hard from S.W. last night, but this morning moderate. Got away & with fair wind, & fine though Showery weather, anchored in Beaver Hr, opposite Fort Rupert before dark. Mr Hunt,[410] in charge here for H.B. Co, came off in canoe with a bag of mail matter, very welcome, containing besides many papers, home letters with good news.

[verso] *Tribes of Indians* speaking closely allied languages, & which may be grouped together under the name of the *Kwa-kuhl Nation*.[411]

Name of tribe	Present Chief	Country
Kō-sk-mo	Kwa-h-la	Greater part of Quatsino Inlet of Chart.
Kwat-zi-no	Ow-t	Forward Inlet
N-wittai	Kow-mād-a-Kwa (or 'Cheap')	Hope &c. Islands & extreme North of Vancouver I.
Kwā-Kuhl	ō-ut	Fort Rupert
Nm-Kish	Kla-sho-ti-awl-ish	Nimpkish of Charts
Li-kwil-tah (=Ukulta)	Yai-ko-tl-is {or wa-mish}	Cape Mudge
Mam-il-i-li-a-ka	Ni-kē-dzi	<Near Nimpkish (on mainland?)> {entrance to Knights Inlet}
Kla-wi-tsush	Hum-tzi-ti-kum-a	<near Alert Bay on Mainland> {Lower end of Clio Channel Turnour I.}

{also these villages}
Nuk-wul-tuh. Mouth of Seymour Inlet / Tsa-wulti-ē-nūh Wakeman Sd.
Tan-uh-tuh Head of Knights Inlet / Met-ul-pai. Havanna Channel {Kwi-'ha on or near Valdez I.[412]
W-wē-kum. Inlet off Nodales Channel.}

Douglas fir. In coming from the north find this tree about the extremity of Vancouver I, though not in great abundance. About entrance to Quatsino Inlet, very few, but abundant, forming extensive groves on upper reaches of the Inlet. Similar distribution obtains, I believe, elsewhere along west coast of Island.

Indians living at Fort Rupert, or Calling it home probably do not exceed 200 in number, according to Messrs Hunt (HB Co) & Rev.

Hall.⁴¹³ They appear to be a dirty, ugly, & degraded lot, not better than those of Koskimo, & infinitely worse looking than the Haidas, Chinseyans, or any of the Northern Indians we have seen.

Coal. Indian reported to Mr Hunt. (HB Co. Fort Rupert) that he found seam of coal about 2 feet thick on a river running into Hardy bay.⁴¹⁴ The locality can be reached in a long summer day when the river low. at present season (oct.) would take two days travel to reach the place. Indians says about as near Koskimo as here. [end verso]

Oct. 2. Morning examined roucks about harbour, collected fossils &c. Afternoon took a couple of photos.⁴¹⁵ Had a talk with Mr Hunt, & Mr Hall, the latter a missionary here, but not yet well acquainted with the Indians of this locality. A fine day but some very heavy showers, Coming over from the westward. The 'Otter' expected on way South at any moment, wrote note to Father, leaving it in charge of H.B.

G.M. DAWSON TO J.W. DAWSON, 2 OCTOBER 1878, FORT RUPERT

I have to acknowledge several letters from you, & quite a *heap* of 'Natures' &c. which I found waiting me here yesterday evening. This is the second mail I have got this season, & brings Montreal dates up to Aug 29! I found here also a letter from Rankine at Victoria, announcing his safe arrival there & intended immediate departure for Montreal. He is by this time I hope Safe at home & hard at work.⁴¹⁶ This note is written merely to announce my arrival here, & I shall leave it for the steamer Otter, which is daily expected, to pick up when she comes along.

I have been round at Quatsino Sound on the West Coast of V.I., & having had rather rough disagreeable weather, leading to much delay & lying at Anchor, am not sorry <....> to get back to the inside again.

The weather now appears settled & likely to be fine for a while, though heavy & almost continuous rains *may* now set in almost any day. Tomorrow I hope to leave here, & after giving a day or two to the examination of the coast between here & Alert Bay, for the purpose of approximately defining the nature & area of the coal-basin,⁴¹⁷ shall make the best of my way to Comox, where I may have to stay a day or two to look at the Baynes Sound Coal Mine,⁴¹⁸ where they have got into some trouble with faults &c. After this Victoria, which *may* be reached in about ten days, with favourable winds, & should this happen I shall try to get things in train to leave Victoria by steamer Sailing on the 20th of this month. These are the plans for the conclusion of the summer's work.

I am anxious to get back as soon as possible to Montreal as the unfinished state of my last years report demands attention <before> to Avoid delay when the rest of the matter goes to press.[419]

Mr Selwyn[420] has not favoured me with a line <since> either on the Subject of the report printed before I left, or that which is to come. It is to be supposed he is very busy & probably regards me as beyond Mail Carriers.

Please congratulate William for me on the success of his essay, & brilliant achievements at the Ecolle. I hope he may succeed in the Toronto Matter,[421] but feel that non-success should be no Cause of surprise, for notwithstanding Capability, he is very young for so responsible a post, where youth is a real disadvantage.

Oct. 3. Up before daylight, breakfasted, & off early with Charley & Chimseyan Johnny in boat. Schooner to follow joining us if possible in McNeil Harbour.[422] Examined coast, landing at many places. Lunched at Su-quash & finally, no sign of schooner appearing, camped in McNeil Hr. A very fine warm bright day with light airs only. Several Indian canoes passed us during the day. Saw also a couple of whites in a boat, fishing salmon with seine for West Huston. They Caught in one haul today about 100 fish. Got a good fire going & find our camp a comfortable one, though weather chilly.

Oct. 4. Off in good time, intending to Coast the South eastern part of Malcolm Island, & then return to Alert Bay to Meet schooner. Found a strong head wind blowing. Ran across to Haddington I. with sail. While examining rocks there saw a young deer on beach, but before rifle obtained it had walked quietly into the woods. Followed it, & catching sight had a shot at it at some distance & in a rather dark place. Missed, as the deer ran away. Followed it again & saw it once more but 'on the jump'. Ran across from Haddington to Malcolm I. but find wind so strong that Could not proceed along shore against it. Waited remainder of day behind a point, reading several 'Witnesses'[423] & dodging the very pungent smoke of a fire, which blew 'every-way. Seeing the schooner beating up, made sail & ran down to her. Beat into Alert bay in rain & squally wind, anchoring at 8.30 P.m.

Oct. 5. Morning Calm. Went in boat to visit bluff at entrance to bay, which turned out to be, not sandstone, but grey well strat. clay. Went through the Indian houses, which built in the style sketched on a former page,[424] not So elaborate as the old Haida style. Some of the ridge-poles, however of great size, & the houses also large, several families (related) occupying the different corners.[425] Saw some very large wooden dishes,[426] not unlike those sketched in old Haida house at North Island, but deeper in proportion. Other wooden dishes of various shapes, [*illus.*]

not precisely like those of Haidas, but similar. The commonest form here & at Quatsino appears to be this [*illus.*] cut from solid block. Saw two large dishes cut from solid, & standing nearly 2 feet high, in this design, the figures representing indians clasping the vessel. [*Illus.*] Another about 4' long was like a large spoon, the end of the handle bent round, & forming a bird's head, which holds a frog (looking also toward the bowl) in its beak. [*Illus.*] Singing going on early this morning, in connection with a potlatch, given by one of the chiefs. Went afterwards into the house, & saw the gentleman counting over & arranging the distribution of blankets in the presence of several of the elders of the tribe. A number of young women employed kneading up dough for bread, which doubtless to form a part of the giving away, which was to be continued, with dancing, in the evening. The song of these people almost exactly in tune (?) & intonation like that heard among the Haidas during a dance.

Westerly wind springing up at noon, got away & made some progress before it died away toward sundown.

Oct. 6. A wasted day. Floated about becalmed in the channel, & eventually carried back beyond our morning position. Anchored in a small cove. Shot a couple of ducks as they swam round the schooner. Heavy rain, which came down as snow not very far up, & when the clouds lifted saw snow on the trees less than 2000' above us on the mountains

Oct. 7. Got away with morning tide, little wind till noon, then light fair wind [in?] Sundown. Then nearly calm. Made some progress, being now opposite entrance to Havanna Channel. Reading & writing.

Oct. 8. Made with the flood, but against a strong head wind, as far as the west end of Thurlow Island. Got into a little bay[427] there, & remained all night.

Reading & writing, though with such slow progress almost too much annoyed to settle down to anything solid

Oct. 9. Off with the flood, & beating up as far as Pender Islands,[428] against an easterly wind. The ebb setting in, the wind fell, & after some trouble we manage to get to an anchorage nearly opposite the islands.

Oct. 10. Anchored this morning for a few hours, during the ebb, & got off again with the flood, a strong south easterly wind blowing up the passage, & raising a tumbling sea. Beat down to Plumper bay, anchoring about 2 P.m., there being too much wind & tide to go through the narrows.[429] Evg. cleared, & became fine & calm. Reading & writing.

Oct. 11. Off about 10 A.m. before the ebb had quite finished running, & got through the narrows almost at slack water. Beat down the passage[430] to Cape Mudge against a S.E. wind, which left us, off the cape, to be tumbled about Some hours in a tide rip. Now (7 P.m.) a light N.W. wind pushing us along gaily. A fine bright night.

Arbutus[431] Saw the first of these trees on the rocks in the narrows. Saw a deer in the woods, at the narrows, but before the rifle could be got, it went off, quietly concealing itself.

Oct. 12. A head wind this morning, but beat down to Harwood Island, which I wished to visit in conformity with request from Mr Sproat[432] for information on it & neighbouring islands. Went off in boat, coasting the west shore. Examined banks of fine strat. sand, & procured sample of soil. Ran across to Comox with a good breeze, tying up at the wharf about 4 P.m. Found no letters, but got hold of a couple of papers with news to Oct. 2., from Victoria. Reading & writing note on islands for Mr Sproat.

Oct. 13. All day occupied in getting down to Baynes Sound Mine wharf, in Fanny Bay, a distance of ten miles. A fine day, but light variable winds. Called in on a Mr Watt,[433] who lives on Denman I., opposite the mine, & is in charge of it, now that all work suspended. Arrange with him to accompany me up to mine tomorrow, to show me the various localities. Glimpses of magnificent mountains, heavily snow clad through the lower range near the coast. These belong to a range marked as 6000 to 7000 feet on the chart, of which Mt Albert Edward is the northern summit. The lower tree-clad mountains already covered with some new snow, Watt says now as far down as it came at any time last winter, which was a remarkably mild one.

Oct. 14. Off early & occupied most of day examining rocks along the railway & about the mines at Baynes Sd. On returning to the coast, took a couple of views[434] of the wharf, & then, there being a strong fair wind got away. Had expected that two days might be required here, but found one sufficient to see everything essential. A fine clear day, but chilly. No rain! Where the rail way to the mine follows parallel to the shore. A cutting has been made in the face of a narrow flat between it & the beach. Here for several hundred yards a great thickness of Indian shell heaps has been exposed. From 12 to 15 feet seen in some of excavations, without bottom being show. Clams, oysters, (small) mussels, sea-eggs &c. form alternating layers or are mixed together. Some layers Calcined, & many burnt stones scattered through the mass. Appears as though village must have been here, but very long ago, as forest trees several hundred years old grow on the upper layers.[435] More or less Earth mixed with the shells in some layers, & occasionally layers of Earth & gravel with few shells. At End nearest wharf the Shelly deposits are interlocked or 'spliced' with the more usual clayey gravels. The explanation of this is that the Soil has been gradually washing down the Slope, in large quantities at certain times, & this has given the spurious appearance of true subaqueous stratification. The structure diagrammatically as below in front view, conjecturally as shown in Cross section – [*Illus.*]

Oct. 15. Got into Nanaimo early this morning. Went about a little business in the town, & then up to find Mr Bryden,[436] at the Coal mine. Waited long for Mr B, he being underground, but at last finding him, accompanied him to his house, & then walked out to Chase River mine[437] & spent a couple of hours underground examining it. Road back on the engine & returned to schooner, to find all hands but Charley missing. By the time that they were again collected breeze ahead & very light, put off departure till morning. Went up to Hospital to Call on Mr Landale[438] for information about Koskimo, but found him able to give very little. Reading &c. in evening. Had a call from Mr Sutton,[439] who lately carried on assaying in Victoria, now running saw mill at Cowitchin.[440]

Without counting extremes, Mr Bryden estimates the Douglas Seam in Chase River Mine to range from 20' to 2' in thickness. Averages about 5' of good coal. Scarcely any shaly partings here, though not so infrequent in old workings which on same seam.[441] Small films of calcite but no visible pyrites. Roof & floor of blackish shale, with coarse sandstone or rather fine conglomerate both above & below. The conglom. forms roof & floor without shale in the old mine. Seam locally variable, but with remarkable general, workable, constancy. Strike very variable & dips undulating causing adits to wind. Dips away first at moderate angle, (about 15 degrees say) then suddenly pitches down without break at <300>' 30 or in places a little more then begins to flatten out again further down on dip. Slope now down over 400 yards. Very little water, that below the adit (over 300' below to bottom of mine) being pumped out by force-pump near bottom of workings supplied with steam from Surface. Ventilation by furnace, at bottom of small shaft heightened above by chimney. Very little gass. No safety lamps used except on tours of inspection by *firemen.* Coal blasted out with powder. Several faults already determined in workings, but of Small throw & coal usually easily recovered. Hauling done by wire rope, drum, engine of 90 horse power steam from 4 long round-ended boilers. Roof & floor firm, & comparatively little timbering needed. Double drifts run in along strike, passage & airway, openings between at intervals, but all of those except a few filled up after passages go ahead of them. [Bards?] opened up to rise, & coal cut out in chambers. Pillars for most part taken away before chambers abandoned. Coal brought down to levels below by back balance, fulls hauling up empties. Mr Bryden does not think any evidence of roots in clay below coal, & fossil plants not find in that above. Believes coal did not grow where now is, but unless so, almost impossible to account for its regularity & purity. [*Illus.*]

Oct. 16. A wet disagreeable morning, with strong head wind. Off at 9.30, reaching Dodd narrows just in time to get through with last of

Ebb. Beating down all rest of day making fair progress. Now (9 P.m.) abreast Narrow island.[442] Reading, packing specimens, & writing.

Oct. 17. Made a considerable advance by steady beating during the night, against a strong head wind, being this morning nearly abreast the entrance to Saanich Inlet. Continued beating all day, the wind falling in the afternoon & leaving up practically becalmed near Trial Island about dark. Drifted on with the ebb tide, however, & finally got a little air to push us into Victoria Harbour. Came up to the wharf from Laurel Pt, with the boat towing ahead. Made fast at 11 P.m. Taking my sponge brush & comb I landed & went up to Driard House, where find all awake. Not Sorry for the present to conclude the Chapter of my <....> marine experiences, Secure a good room & look with admiration on a bed with Clean sheets!

[Dawson's journal closed on 17 October and he reverted to his small notebook. While much is routine, the remainder of his stay in Victoria is sufficiently interesting to be reproduced.]

Victoria Oct. 18. Got all stores specimens &c. out of schooner, having previously made arrangements to have them placed in one of H.B. stores on wharf. Settled accounts with Williams Charley & Indian Johnny. Attended to various other matters of business. Evg. dined at Judge Greys,[443] Meeting a Dr & Mrs Hanington[444] there.

Oct. 19. To Muirheads[445] to get carpenter send down to take flooring &c. out of schooner. Secured a number of empty boxes for specimens & had them sent down to store. Packing specimens all morning with Charley. Store closing at 1 P.m. – being Saturday – prevented work after lunch. Interviewed several people on mining &c. & took a turn in Chinese quarter to look for China. Dined at Judge Crease's.[446] A bleak wet stormy day.

Settled with Sabiston today, & glad to get rid of him.

Oct. 20. Down to Schooner to get Sabiston to sign a new receipt, the old one being defective. Dr Tolmie[447] arriving at 10 A.m. went with him to his house, & remained there conversing & working chiefly on Indian vocabularies[448] most of day. Dinner about 4 P.m. Drove back to Town before 6. Had a cup of tea & set to work on correspondence &c. A fine but cool day.

G.M. DAWSON TO J.W. DAWSON, 20 OCTOBER 1878, VICTORIA

I have now been in Victoria three days, having arrived at the wharf, in the 'Wanderer' at about Eleven o'clock on Thursday Night. Everything is now out of the Schooner, & much of my stuff packed, men paid off &

affairs generally in progress towards an end. I shall not be able to leave by tomorrow's steamer, however, – by which this letter goes, – but have not yet decided whether to seek an intermediate boat at Portland, or wait for that of the 30th from this place. The latter date seems a long way off now, but I have a number of things to look after, when my Packing is done, which may detain me too long to allow me to <....> Save anything in time by going *via*. Portland. The latest date from Montreal is your letter of Sept. 28th, in which you announce receipt of letters, among others one from Rankine from Portland. R. has now no doubt been home some time.

The getting from Fort Rupert to this place has been a longer operation than I thought probable, owing to the prevailing South-easterly winds of this season. The weather is now very uncertain & beginning to be very wet, so that the conclusion of field work is not to be regretted.

The railway office has been moved over to New Westminster, & I was told last night that some mail matter is over there belonging to me. These may be letters which have been addressed to care of C.P. Ry, & account for Certain lacunae in home correspondence which appeared to be indicated by letters found here. I have to acknowledge a long letter from William, but shall not attempt to answer it at this time, as there is nothing in particular to write about.

Enclosed you will find a little bill, which reached me at Pt. Simpson, but which seems to be intended for you, as it is for extras of a Paper on Canadian Phosphates.[449]

When I decide on route, I Shall write again, but will not telegraph unless something of special importance turns up.

Oct. 21. Packing specimens &c. & making list of articles left. Removed boxes from room in Custom House to H.B. store. Saw Mr Fawcett[450] & Deans[451] on QCI anthracite. Evg. spent with Mr Woodcock getting notes on Gold Hr & Omineca country. A steady downpour of cold rain nearly all day.

Oct. 22. Down to wharf to see about carpenter closing boxes. With Mr McKay (H B Co) examining specimens. Looking over specimens & plans of Hows Sd Copper Mine[452] with Roscoe[453] & others. With Messrs Finlayson[454] & Heistermann[455] examining plans &c of Baynes Sd. Mine. At Lands & Works office looking for plans for map. Call at McDonalds but find no one at home. Looking over plans of Skidegate Coal mine at Fawcetts.

Oct. 23. At wharf to superintend carpenter closing & addressing boxes. At Lands & Works office getting trace of Okanagan & Spallumsheen region.

After lunch back to L & W office to look over maps & plans there. Find a number which it will be well to have traces of. Call at Judge Greys & Governor Richards, & stay to dinner with latter. A mail of papers this evg. from Montreal.

Oct. 24. At Lands & Works making tracings. P.m. with Dr Tolmie & Haida Mills getting Haida vocabulary. Wrote Hunter[456] & Robson[457] on Survey from Duck & Pringles[458] to Grande Prairie.[459] Authorized former if possible to get work done under $120 to proceed if he can find surveyor. Asked Robson for transport from CP Ry.

Evg. writing to West Huson & T.H. Huxley.[460]

G.M. DAWSON TO MARGARET DAWSON, 24 OCTOBER 1878, VICTORIA

Your welcome letter of October 8th, reached me a couple of days ago, & I am glad to know under such late date that all are well at home, but sorry to hear that William has been disappointed in the Toronto Affair, especially as so much trouble of one sort or another has been taken about it.

I have decided not to leave here till the regular boat sails next Wednesday, finding that by going *via* Portland a day or two sooner I cannot make sure of gaining anything, while by staying here I can get some tracings made in the Lands & Works office & look after other matters.

You ask me in your letters to telegraph on my arrival at Victoria, but as I had already been here some days when I received the letter, & a written notification of my arrival was on the road I thought I might dispense with the Ceremony. Once on the way I shall loose no time in getting to Montreal, & if in good luck may arrive about Novr. 11.

The weather is fine here now, but not really settled. It is beginning to be cool, with occasional frosts at night, & all the old inhabitants foretell an early & severe winter – for this place.

Rankines description of Victoria contributed to the *Witness* has been reprinted in one of the papers here,[461] but owing to the signature & place from which he dates everyone attributes it to me, & wonders why I have been four years here & only found all that out now!

I may write again before leaving, though it is scarcely probable that letters later than this <can> will reach Montreal before I do.

Please forgive this Scrawl

Oct. 25. Up at 6 Am & down to Esquimalt with Capt Goudy[462] to see Bark Hazelhurst, which about to proceed to China with Chinese passengers, & freight. Breakfast on board & on return go out to Dr Tolmie's. At work on Indian matters till 5 P.m. Drove back with Dr T.

Oct. 26. At Lands & Works office till after noon tracing maps. After lunch determined sextant index error. About town on various little matters. Dined at Duponts' with Mr & Mrs Fellows,[463] Mr Burns[464] & Miss Martineau.[465]

Oct. 27. Going over things in hotel, packing & making list of what left. P.m. for a walk to Beacon Hill. Evening going over Indian matters & getting list of questions for Mills. Saw Capt. D.E. Sturt[466] of Texada Island Marble quarry.[467] He took me to see samples of the marble in a stone yard on Government Street. Says there is *coal on mainland* inside Texada Island. Sandstones extend about 9 miles along shore S.E. of the Indian Village. Coal seam a mile or so inland & where he saw it 1 foot thick burns well.

Oct. 28. Called on Archdeacon Wright,[468] having understood that he wished to ask about some Indian matters.

Booked for *Dakota* on Wednesday. Got boxes from H.B. taken down to wharf, consular certificates, Bills of lading &c. for them. Found box of specimens from Montreal in bonded warehouse. Paid bills & transacted various business about town. Met Dr Tolmie after lunch to go over Haida vocabulary but Indian Mills did not turn up. Went down to Indian quarter to look for him, but unsuccessfully.

Met Mr Fawcett to discuss Q.C. Anthracite mine.

Evg. writing notes & letters wrote Bowman,[469] W Duncan & Rev Mr Collinson.

reading.

Oct. 29. Paid for ticket to San Francisco. Paid various bills. Got boxes down to HB. store, & Blankets returned to C.P. room in Custom House Building. Called at Lands & Works Office. Lunched at McDonalds. P.m. Saw Dr Ash,[470] Spencer[471] on photos, & Dr. Tolmie.

Dined with At Wards,[472] with Mr [Morglan].[473] Packed up everything. To bed at 12.

Oct. 30. Paid hotel bill Saw Hibben[474] on Reports. Left for Esquimalt in Bowmans buggy[475] at 10.30. Left in S.S. Dakota punctually at noon, passing inside Race Rock light, & abreast Cape Flattery about dark. Magnificent view of Mt. Baker & the serried peaks of the coast range. Clear sunny day, with beautiful glow at sunset on Cape Flattery & Vancouver shore. A slim passenger list & plenty of room everywhere.

[Dawson arrived in San Francisco on the morning of 2 November and departed on the Central Pacific Railway the next day, changing at Council Bluffs for Chicago. Travelling over Detroit, Stratford, and Toronto, he reached Montreal on 10 November 1878.]

On the Haida Indians of the Queen Charlotte Islands

The following account of the Haida Indians is chiefly the result of personal observations during the portion of the summer of 1878 spent in the Queen Charlotte Islands, prosecuted during moments not occupied by the geological and geographical work of the expedition, at the camp fire in the evening, or on days of storm when it was impossible to be at work along the coast. I am also indebted to the Rev. Mr. Collison, of the Church Missionary Society, for various items of information, and largely to Dr. W.F. Tolmie, of Victoria, for comparative notes on the Tshimsians. Mr. J.G. Swan has published a brief notice of the Haidas in the Smithsonian Contributions to Knowledge, vol. xxi, no. 267, 1876. This may be consulted with advantage on some points, more particularly on the nature of the tattoo marks of these people. The present memoir is, however, I believe the first detailed account of the Haidas which has been given.

The Haida nation appears to be one of the best defined groups of tribes on the north-west coast. Its various divisions or bands differ scarcely at all in customs, and speak closely related dialects of the same language. They have been from the earliest times constantly in the habit of making long canoe voyages, and taking into account the ease with which all parts of their country can be reached by water, it would indeed be difficult to explain the slight differences in dialect which are found to exist, but for the knowledge that in former times they carried on, at least occasionally, intertribal wars; besides constituting themselves, by their warlike foreign expeditions and the difficulty of pursuing them to their retreats, one of the most generally dreaded peoples of the coast, from Sitka to Vancouver Island. This warfare, however, partook of the barbarous character of that of the other American aborigines, and consisted more frequently in the surprise and massacre of helpless parties, even including old people and women, than in actual prolonged conflict.

The original territory of the Haidas, as far as tradition carries us back, is the well-defined group of islands called by Captain Dixon in 1787 the Queen Charlotte Islands, but which the people themselves call *Hai-da-kwē-a*.[1] These islands lie between the latitudes of 51° 55' and 54° 15', with an extreme length of about 190 miles. They are separated by waters of considerable width from the mainland to the east and from the southern extremity of the territory of Alaska to the north. At the present day, however, people of the Haida stock, and closely related in every way to the tribes of the northern end of the Queen Charlotte Islands, occupy also a portion of the coast of the southern islands of Alaska, being the south end of the Prince of Wales Archipelago, from Clarence Strait westward, together with Forrester's Island.

It has been supposed that from the large islands adjacent to the mainland the Queen Charlotte Islands have been peopled, but this is not the case, for the traditionary account is still found among the natives of internecine wars as a result of which a portion of the Haidas of the northern part of the Queen Charlotte Islands were driven to seek new homes on the Prince of Wales group. Their story is borne out by other circumstances, and the date of the migration cannot be more than 150 years ago. These Haidas living beyond the Queen Charlotte group are generally known collectively as *Kai-ga-ni*, which name is also among the Indians applied to the country they inhabit.

Frequently, among tribes pretty closely related in language, the process of differentiation has gone so far that neighbouring peoples disclaim any community of race, though on comparing their vocabularies their national identity becomes apparent. This is not the case, however, among the Haidas, who speak of all the people of their nationality as Haida, adding when necessary the name of the region inhabited by the tribe. A comparison of the Haida language with those of the other tribes of the coast shows very few points of resemblance.

PHYSICAL PECULIARITIES AND DRESS

Physically, the various tribes of the north-west coast differ to some extent, so that a practised eye may distinguish between them, but the differences are slight as compared with those obtaining between the coast tribes generally, and those of the interior of British Columbia. The Haidas are, however, markedly fairer skinned than most of the coast tribes, and possess somewhat finer features. In the coarseness of the mouth, width and prominence of the cheek bones, and somewhat disproportionately large size of the head as compared with the body, the main departures from ideal symmetry are to be found. The body is also not infrequently long and large as compared with the legs, a circumstance doubtless brought about by the constant occupation of these people in canoes and the infrequency of their land excursions. The hair is black and coarse, and only in the case of 'medicine men' have I observed it to be allowed to grow long in the male sex. A scanty moustache and beard sometimes clothe the upper lip and chin, generally in the case of old people who have given up the habit of eradicating the hair as it grows. In some instances, and these more numerous than in the other coast tribes, both men and women of prepossessing appearance, and with features of considerable regularity as measured by European standards, occur. The average physiognomy of the Haida shows more evidence of intelligence and quickness than that of most of the coast tribes, an appearance not belied on more careful investigation. I have not been able to discern in their appearance anything of that exceptional fierceness said to be characteristic of them by the earlier voyagers, and can only suppose that these statements may have arisen from the more elaborate character of their armament and dress, and the liberal application of pigments to the skin. Many of the Haidas are said to be strong and dexterous swimmers, but I have never seen them exercising the art, which may probably be reserved for occasions of necessity. They are not long-lived, though grey-haired men and women may occasionally be seen. Pulmonary diseases accompanied by spitting of blood, and blindness generally caused by a species of opthalmia, are not uncommon; and other diseases incident to a life of exposure tend to reduce the term of life, as they do among all the aborigines of the continent. Besides these, however, and much more fatal, are diseases introduced among them since contact with the whites. Great numbers of the Haidas, with all the other tribes of the coast, have been cut off by small-pox, both during their periodical visits to Victoria and after their return to their native islands. This disease is with them almost certainly fatal, and I could learn of a single instance only in which recovery had occurred. Owing to the complete demoralization of the Haidas since contact with the whites, and their practice of resorting to Victoria and other places, where they maintain themselves by shameless prostitution, venereal diseases are extremely common and destructive.

In dress the Haidas, like other Indians, have adopted, so far as their means enable them, the customs of the whites, though their costume as a rule might be considered rather scanty, and some of the older people use scarcely anything but a blanket as a protection from the elements. The blanket with these people has replaced the 'robes of sea-otter skins' which so much pleased the eyes of the early traders. In Dixon's narrative[2] the sea-otter 'cloaks' are said to 'generally contain three good sea-otter skins, one of which is cut in two pieces; afterwards they are neatly sewed together so as to form a square, and are loosely tied about the shoulders with small leather strings fastened on each side.' The women's dress is more particularly described on another page in the following terms: – 'She was neatly dressed after their fashion. Her under garment, which was made of fine tanned leather, sat close to her body, and reached from her neck to the calf of her leg; her cloak or upper garment was rather coarser, and sat loose like a petticoat, and tied with leather strings.'

These extracts both refer particularly to the Haidas, but in the general account of the natives of this part of the north-west coast, the dress of the people is more minutely described in the following paragraph: – 'In their dress there is little variety; the men generally wearing coats (such as I have already described) made of such skins as fancy suggests or their success in hunting furnishes them with, and sometimes the loose cloak thrown over the shoulders and tied with small leather strings. Besides this, some of the more civilized sort, particularly those in Cook's River, wear a small piece of fur tied round the waist when the heat of the day causes them to throw their coat aside or they are disposed to sell it. The dress of the women differs in some respects from that of the men. Their under garment is made of fine tanned leather, and covers the body from the neck to the ankle, being tied in different parts to make it fit close; over this is tied a piece of tanned leather like an apron, and which reaches no higher than the waist. The upper garment is made in much the same manner as the men's coats, and generally of tanned leather, the women not caring to wear furs, as they were always unwilling to be stripped of their garments, which, should they happen to be worth purchasing, their husbands always insisted on their being sold. Indeed, the deportment of the women in general was decent, modest and becoming.'

In former days a sort of armour was worn, consisting of split sticks arranged in parallel order and combined with the stronger parts of the hide of the sea-lion. None of these suits can now, however, be found. A cloak or blanket very much prized by the Haidas and called *naxin* is obtained in trade from the Tshimsians. It is shaped somewhat like a shawl, with a blunt point behind, and surrounded by a deep and thick fringe of twisted wool. Finely shred cedar bark is used as a basis or warp, on which the wool of the mountain goat is worked in. The cloaks are made in many small separate pieces, which are afterwards artfully sewn together. The colours of wool used are white, yellow, black and brown, and the pattern bears a relation to the totem, so that an Indian can tell to what totem the cloak belongs. These cloaks or blankets are valued at about $30. They are used specially in dancing, and then in conjunction with a peculiar head-dress, which consists of a small wooden mask ornamented with mother-of-pearl. This stands up from the forehead, and is attached to a piece fitting over the head, ornamented with feathers, &c. and behind supporting a strip of cloth about two feet wide, which hangs down to the feet, and is covered with skins of the ermine. The cloaks are described by the chronicler of Dixon's voyage as a 'kind of variegated blanket or cloak, something like our horse-cloths; they do not appear to be wove, but made entirely by hand, and are neatly finished. I imagine that those cloaks are made of wool collected from the skins of beasts killed in the chase; they are held in great estimation, and only wore on extraordinary occasions.'

Shred cedar bark, twisted into a turban, and stained dull red with the juice of the bark of the alder, is frequently worn about the head, more, however, as an ornament than a covering, and apparently without any peculiar significance among the Haidas, though with the Tshimsians and the Indians of Millbank Sound it is only worn on occasions of religious ceremony, and it would be considered improper at other times.

Feathers, buttons, beads, portions of the shell of the Haliotis, with the orange-coloured bill of the puffin, are used as ornaments, strung together or sewn on the clothes. The Dentalium shell was formerly prized and frequently worn, but has now almost disappeared.

Painting is frequently practised, but is generally applied to the face only. Vermillion is the favourite pigment, and is usually – at least at the present day – rubbed on with little regard to symmetry or pattern. Blue and black pigments are also used, but I have not observed in any case the same care and taste in applying the paint to form a symmetrical design as is frequently seen among the Indians east of the Rocky Mountains. The face is almost always painted for a dance, and when – as very often happens – dances recur on occasions of ceremony for several nights, no care is taken to remove the pigment, and most of the people may be seen going about during the day with much of it still adhering to their faces. To prevent unpleasant effects from the sun in hot weather, especially when travelling, the face is frequently first rubbed with fat, and then with a dark brownish powder made by roasting in the fire the woody fungus found on the bark of trees, and afterwards grinding it between stones. This soon becomes nearly black, and resembles dried blood. A mixture of spruce-gum and grease, also of a dark colour, is used to protect the face in cold weather, while those in mourning frequently apply grease and charcoal to the face.

Bracelets beaten out of silver coins are very generally worn by the women, who often carry several on each arm. The custom of wearing several or many polished copper rings on the ankles and arms was formerly common among the Haidas and Tshimsians. Those for the ankles were round in section, those for the arms flat on the inner side. In Dixon's narrative 'large circular wreaths of copper' are spoken of as being frequently worn, both at Norfolk Sound and in the Queen Charlotte Islands. They 'did not appear to be foreign manufacture, but twisted into that shape by the natives themselves to wear as an ornament about the neck.'

Tattooing is universally practised, or rather was so till within the last few years, for it is noticeable that many of the children are now being allowed to grow up without it. The front of each leg above the ankle and the back of each arm above the wrist are the places generally chosen, though the breast is also frequently covered with a design. The patterns are carefully and symmetrically drawn, of the usual bluish colour produced by the introduction of charcoal into punctures in the skin. In one instance, however, a red pigment had also been employed. The designs are often hereditary, and represent the totem crest of the bearer, in the usual conventional style adopted by the coast Indians in their drawings. I have never observed any tattooing to extend to the face, where it is commonly found among the Tinneh people of the interior, in the form of lines radiating from the corners of the mouth, on the chin or forehead.

Till quite lately the females among the Haidas all wore labrets in the lower lip. Dixon particularly notes this as being the case, though in Norfolk Sound it was only practised by women of rank. Dixon further gives an admirable illustration of the Haida labret in the plate facing page 226 of his volume, already several times referred to. A small aperture first made is gradually enlarged by the insertion of lip-pieces of ever-increasing size, till the lower lip becomes a mere circle of flesh stretched round the periphery of a flat or concave-sided labret of wood or bone, which projects at right angles to the plane of the face. One obtained by Dixon was found by him to measure $3\frac{7}{8}$ inches long by $2\frac{5}{8}$ broad, which is larger than any I have seen. Only among the old women can this monstrosity be now found in its original form. Many middle-aged females have a small aperture in the lip, through which a little beaten-silver tube of the size of a quill is thrust, projecting from the face about a quarter of an inch. The younger women have not even this remnant of the old custom.

The piercing of the lip was the occasion of a ceremony and giving away of property. During the operation the aunt of the child must hold her. The shape of the Haida lip-piece or *stai-e* was oval. Among the Tshimsians it was more elongated, and with the Stickeen women nearly circular. It was also formerly the custom to pierce the ears in several places. Three perforations in each ear were usual among common people, but chiefs or those of importance had five or six. These held little ornaments formed of plates of haliotis shell backed with thin sheet copper, or the small sharp teeth of the fin whale. This custom obtains also among the Tshimsians and Stickeen Indians, and the Chiefs Callicum and Maquilla of Nootka Sound, Vancouver Island, are represented with the same adornment in Meares' engraving of them.

The septum of the nose is generally perforated in both males and females, and was formerly made to sustain a pendant of haliotis shell or a silver ring, though it is not now used in this way. No process of distortion of the head or other parts of the body is practised among the Haidas.

FOOD

Like most of the tribes of the coast, the Haidas live principally on fish. The halibut and salmon are chiefly depended on. A complete list of the articles used by them as food would, however, indeed be a long one, as few organic substances not absolutely indigestible would be omitted.

The halibut fishery is systematically pursued, and the main villages are so situated as to be within easy reach of the banks along the open coast on which the fish abounds. The halibut is found in great numbers in all suitable localities from Cape Flattery northward, but is perhaps nowhere finer, more abundant, and more easily caught than in the vicinity of the Queen Charlotte Islands. It may be taken in most of the waters at almost any season, though more numerous on certain banks at times well known to the Indians. About Skidegate, however, it is only caught in large numbers during a few months in the spring and early summer. When the fish are most plentiful the Haidas take them in large quantities, fishing with hook and line from their canoes, which are anchored by stones attached to cedar-bark ropes of sufficient length. They still employ either a wooden hook armed with an iron – formerly bone – barb, or a peculiarly curved iron hook of their own manufacture, in preference to the ordinary fish hook. These implements are described with others in treating of the arts of the Haidas.

The halibut brought to the shore are handed over by the men to the women, who, squatted on their haunches, rapidly clean the fish, removing the larger bones, head, fins and tail, and then cutting it into long flakes. These are next hung on the poles of a wooden framework where, without salt – by the sun alone, or sometimes aided by a slow fire beneath the erection – they are dried, and eventually packed away in boxes for future use.

There are no rivers of great size on the islands, but many streams large enough to be known as 'salmon rivers' to the Indians. A run of small red-fleshed salmon occurs about the middle of July up some of the larger streams. These answer no doubt to the fish known on the Fraser River as the suckeye, and much prized. They are, however, in inconsiderable numbers, and not much sought after by the Haidas. About the middle of August a larger species begins to arrive in great numbers, and this run sometimes lasts till January. These fish when they first appear and are still in salt-water are fat and in good condition. They soon begin, however, to become hook-billed, lean and pale-fleshed. They ascend even very small streams when these are in flood with the autumn rains, and being easily caught and large, they constitute the great salmon harvest of the Haidas. They are generally either speared in the estuaries of the streams or trapped in fish-wiers made of split sticks, which are ranged across the brooks. The various 'rivers' are the property of the several families or subdivisions of the tribes, and at the salmon fishing season the inhabitants are scattered from the main villages; each little party camped or living in temporary houses of slight construction in the vicinity of the streams they own.

It is scarcely necessary to particularize at length the other species of fish used as food, comprising all those abundant in the vicinity of the islands. Trout, herring, flounder, rock-cod, &c., constitute minor items in the dietary. The mackerel and cod are found, but not specially sought after by the Indians, and it is not yet known whether at certain seasons and localities they may be sufficiently abundant to attract commercial enterprise. The spawn of the herring is collected on spruce boughs placed at low water on the spawning grounds, dried and stored away in a manner exactly similar to that practised by most of the coast Indians. The pollock is found on the western coast. It is generally caught in deep water with hook and line, and owing to its fatness is much prized. The Haidas of Gold Harbour or Port Kuper make an annual business of catching these fish in the latter part of the summer. They extract the oil from them by boiling in large wooden boxes with hot stones, and then skimming it from the surface. The oil is carefully stored away, and used as a condiment to dried fish or berries, instead of the oolachen grease, which by this tribe of Haidas is not much in request.

Both the Haidas and Tshimsians have the custom of collecting salmon roe, putting it in boxes, and burying these below high-water mark on the beach. When decomposition has taken place to some extent, and the mass has a most noisome odour, it is ready to eat, and is considered a very great luxury. Sometimes a box is uncovered without removing it from the beach, and all sitting round eat the contents. Fatal poisoning has followed this on several occasions. It is attributed to a small worm which is said at times to enter the decomposing mass from the sea. The Haidas also occasionally allowed the heads of salmon and halibut to lie on the beach between high and low water marks till partly decomposed, when they were considered to be much improved.

The dog-fish is very abundant along some parts of the coast, and its fishery is now beginning to be engaged in. The fish is not eaten by the Haidas, but the oil extracted from the liver is readily sold to white traders, and constitutes one of the few remaining articles of legitimate marketable value possessed by the natives. Large sharks abound on the northern and western coasts, and are much feared by the Haidas, who allege that they frequently break their canoes and eat the unfortunate occupants. No instance of this kind is known to me, but they fear to attack these creatures. When, however, one of them is stranded, or found from any cause in a moribund state, they are not slow to take advantage of its condition, and from the liver extract a large quantity of oil. The whale and hair-seal (if it be proper to include these among products of the fisheries) abound in the waters surrounding the islands. I cannot learn that the former were ever systematically pursued as they were by the Makah Indians of Cape Flattery and Ahts of the west coast of Vancouver Island. When, however, by chance one of these comes ashore it is a great prize to the owner of the particular strip of beach on which it may be stranded. The seal is shot or speared, the latter doubtless having been the primitive mode. Both the flesh and blubber are eaten, the Indians comparing the animal on account of its fatness to that – to many of them hypothetical creature – from which pork is derived. They speak of it in the Chinook jargon as *si-wash co-sho*.[3] It is interesting to remark in this connection that most of the Haidas will on no account eat pork, for some reason which I have been unable to determine.

The oyster is not found on the coasts of the Queen Charlotte Islands, though it occurs in some sheltered localities about Vancouver Island. Clams (*Saxidomus squalidus, Cardium Nuttalli*, &c.,) however, abound, with the large horse mussel (*Mytilus Californianus*) which on rocks exposed to the full force of tidal currents attains a great size. These shellfish of course form a portion of the native diet. They are not eaten, however, at all seasons, but during the winter months only. At other times (April to October) they are reputed to be poisonous, and more than once have proved fatal to those eating them. The Indians attribute this to a worm which they say during the summer season inhabits the cavity of the shell. The Tshimsians and other northern tribes also abstain from shell-fish during the summer for the same reason, while those of the southern part of Vancouver Island appear to eat them at all seasons.

Chitons, both the large red species (*Cryptochiton Stelleri*) which sometimes attains a length of eight inches, and the smaller black variety (*Katherina tunicata*), very common everywhere near low-water mark, are favourite articles of diet.

Sea-urchins, the large purple-spined (*Loxechinus purpuratus*) and the small green species (*Euryechinus chlorocentrotus*), are often brought ashore in large quantities, and it is surprising to observe how many of these rather watery creatures an Indian – squatting perhaps on his haunches on the beach – will devour in making a light lunch. A gentle knock on a stone serves to open the shell, when the finger run round the smooth interior brings out the edible parts, consisting chiefly of the more or less mature ova.

A large brown tuburculated holithurian is also eaten, though some of the younger people now profess to eschew these rather unpleasant looking animals.

Oolachen grease, called *tow* is an important and much relished constituent of many of the Haida dishes. The oolachen or candle-fish, (*Thaleichthys pacificus*) from which it is derived, does not occur in the waters surrounding the Queen Charlotte Islands. It is found in some of the inlets on the west coast of Vancouver Island, but is especially abundant at the spawning season, in early spring, in the estuaries of the larger rivers of the mainland, and of these pre-eminently in the Fraser and the Nasse. Like its eastern representative and zoological ally, the capelin, it swarms in the shallow water along the shore, and is easily caught in immense numbers. For the extraction of the oil the fish is generally allowed to partially putrefy, and is then boiled in a mass in wooden boxes, with hot stones. The oil or grease is semi-solid when cold, with a fœtid and rancid smell and taste. From the Nasse fisheries the oil is obtained by barter by the inland tribes of the northern part of British Columbia and by the Haidas. For a box containing somewhat over one hundred pounds of this grease from six to ten 'blankets,' or say from $12 to $20, is paid.

With dried fish, dried or fresh berries, and in fact with food of any description, no condiment is so grateful to the Haida palate as this oolachen grease; and in the absence of farinaceous substances, it doubtless enables the otherwise imperfect food to go further in supplying the wants of the system.

The Haidas are not great hunters. They kill a considerable number of black bears at two seasons of the year, when they are found prowling along the sea shore, but do not follow them far into their mountain fastnesses. In early spring, when the grass along the edges of the woods begins to grow green, with the skunk-cabbage (*Lysichiton Kamtschatense*) and other succulent vegetables, bruin coming out to browse upon the tender shoots may fall a victim to the lurking Indian. Again in autumn, when tempted to the shores and estuaries by the dead and dying salmon, he is apt to get into trouble, and at this season his skin, being in good condition, is of some value.

There is pretty good evidence to show that the wapati occurs on the northern part of Graham Island, but it is very seldom killed. The small deer (*C. Columbianus*) is not found on the islands, nor is the wolf, grizzly bear, mountain sheep or mountain goat. Geese and ducks in vast numbers frequent the country about Masset and Virago Sound in the autumn, and for a time form an important item in the diet of the natives. They now shoot them with the flint-lock trade muskets with which they are generally armed. I have seen a bow, with blunt wooden arrows, also in the canoe, to be used in despatching wounded but still living birds, and thus to save ammunition. Sea-fowl of many kinds are articles of food on occasion, though the gull, the loon and some others are exempt on account of their exceptionally rank flavour. The eggs of sea-birds, and especially those of the large white gull, are collected in great quantity in the early summer. Every lonely and wave-washed rock on which these birds deposit their eggs is known to the natives, who have even these apportioned among the families as hereditary property. The singular rocks extending southward from Cape St. James are frequented by myriads of sea-fowl, and some of them are so abrupt and cliff-surrounded that, lashed by the never-ceasing swell of the Pacific, they remain inaccessible even to the Haidas.

The potato, called *skow-shit* in Haida, introduced by some of the early voyagers, now forms an important part of the food supply. A Skidegate Indian told me that it was first grown at Skidegate, but I do not know how far this statement may be reliable. The greater part of even the flat low lands of these islands is so thickly wooded, and with trees of such great size, that the task of clearing the ground is quite beyond the energy of the Indian. There are places, however, near the shore, where by cutting down and grubbing out small bushes limited garden patches may be made. These are very often spots which have been occupied by Indian houses, and where great quantities of shells and other refuse have accumulated, forming a rich soil. Such spots are utilized as potato gardens, but are generally small and often scattered far away from the main villages, wherever suitable localities can be found. Little attention is paid to the cultivation of the plant, and the variety in use is generally run down so as to yield very small and poor tubers.

Formerly many small roots indigenous to the country, and containing more or less starch, were eagerly sought after, dried and stored away. One of these was a wild lily. No effort is now made to gather these, though a few may be collected where they occur abundantly. The cambium layer of the spruce (*A. Menziesii*) and hemlock (*A. Mertensiana*) is collected, the trees being cut down and barked for the purpose, and is eaten in a fresh or dried state. This substance has a not disagreeable sweet and mucilaginous taste, but also possesses a distinct resinous flavour. It is considered very wholesome. The cambium layer of the scrub pine (*P. contorta*) is not eaten, though this tree is found in some abundance on the west coast of the islands, and on the mainland of British Columbia is barked for this purpose almost exclusively. The growing shoots of the epilobium, heracleum and other plants are eaten when in season. A sea-weed resembling dulce, but which I have only seen in dried cakes, is found, especially in the southern islands, preserved by drying and boiled into a sort of tea or soup.

Berries abound, the most important being the sal-lal (*Gaultheria shallon*), known to the Haidas as *skit-hun*, and crab-apple or *kyxil* (*Pyrus rivularis*). The latter, about one-third of an inch in length and less in width, has much the taste of a sour Siberian crab. It is gathered late in the autumn, and generally boiled and put away in boxes, covered with water, and allowed to remain so till winter, when the berries are sorted, mixed with oolachen grease, and thus made ready for use. The sal-lal berries are eaten fresh in great quantities, and are also dried for use in winter. The strawberry (*Fragaria Chilensis*), flowering raspberry (*Rubus Nutkanus*), current (*Ribes sp.*), *Vaccinium parviflorum*, &c., occur in some places abundantly. The mahonia (*Berberis aquifolium*) is not found. The service-berry (*Amalanchier alnifolia*), so much prized by the Indians of the interior, occurs sparingly, and scarcely seems to ripen its fruit.

Before the introduction of the potato, the only plant cultivated was one which has been described to me as 'Indian tobacco.' There is a mythical tradition concerning the origin of this plant, which is given in another place. Its cultivation is now entirely abandoned except at Cumshewa, where a single old woman continues to grow it, some of the older Indians still relishing it. This I learnt after leaving Cumshewa, and having consequently been unable to ascertain whether the plant is really tobacco or not. It is probable, however, that it is some less potent weed, or its cultivation would not have been so soon given up and high prices paid for imported tobacco. The Haidas used to grow it not only for themselves, but as an article of trade with other neighbouring tribes. To prepare the plant for use it was dried over the fire on a little framework, finely bruised in a stone mortar, and then pressed into cakes. It was not smoked in a pipe, but being mixed with a little lime prepared by burning clam-shells, was chewed or held in the cheek. The stone mortars – elsewhere more fully described – are still to be found stowed away in corners of the houses. They appear to have been used in the preparation of the 'tobacco' only, and though often large enough for the purpose, were certainly not employed to reduce any cereal to the state of meal, as none such were known to the Haidas. It is, therefore, unsafe to conclude from the mere discovery of stone mortars, among other relics, that certain extinct tribes cultivated corn and used it as food. The leaves of the bear-berry or kinnikinick (*Arctostaphylos uva-ursi*) are mixed with tobacco when smoking, to eke out the precious narcotic. These leaves are used for the same purpose by the Indians everywhere over the northern part of the American continent. I have seen on Vancouver Island the leaves of the sal-lal roasted before the fire and mixed with tobacco, and among the Chippeway Indians and others the bark of the red osier dog-wood (*Cornus stolonifera*).

The dog is the only domesticated animal among the Haidas. The original breed is now much disguised by imported strains. The present natives are grey wolfish-looking curs about the size of a coyote.

SOCIAL ORGANISATION

The Haidas, like other tribes inhabiting the coast of British Columbia and its adjacent islands, have permanent villages. The general type of construction of the houses in these is nearly the same among all the tribes, but among the Haidas the buildings are more substantially made, and much more care is given to the accurate fitting together and ornamentation of the edifice than I have elsewhere seen. This may be due in part to the comparatively late date at which the Haidas have come closely in contact with the whites, but probably also indicates an original greater facility in constructive and mechanical processes than is found among the other tribes. This would be fully borne out by their present character in these regards. Especially in the great number, size, and elaborate carving of the symbolical posts, is this superiority shown. Among the Tshimsians at Port Simpson, most of the original carved posts have been cut down as missionary influence spread among the people. At Nawittie (Hope Island), Quatsino Inlet (Vancouver Island) and elsewhere, where the natives are still numerous and have scarcely been reached by missionaries, though similar posts are found, they are small, shabby, and show little of the peculiar grotesque art found so fully developed among the Haidas.

As before mentioned, the permanent villages are generally situated with regard to easy access to the halibut banks and coast fisheries, which occupy a greater proportion of the time of the natives than any other single employment. The villages are thus not infrequently on bleak, exposed, rocky coasts or islands, though generally placed with care, so as to allow of landing in canoes even in stormy weather. The houses may stand on a flat, elevated a few feet above the high-tide mark, and facing seaward on a sandy or gravelly beach, on which canoes can be drawn up. The houses are arranged side by side, either in contact, or with spaces of greater or less width between them. A space is left between the fronts of the houses and edge of the bank, which serves for a street, and also for the erection of the various carved posts, and for temporary fish-drying stages &c. Here also, any canoes are placed which it is not desired to use for some time, and are carefully covered with matting and boughs to protect them from the sun, by which they might be warped or cracked. As a rough average, it may be stated that there are at least two carved posts for each house, and these, when the village is first seen from a distance, give it the aspect of a patch of burnt forest with bare, bristling tree-stems. The houses themselves are not painted, and soon assume a uniform inconspicuous grey colour, or become green or overgrown with moss and weeds, owing to the dampness of the climate. The cloud of smoke generally hovering over the village in calm weather, may serve to identify it. Two rows of houses are occasionally formed, where the area selected is contracted. No special arrangement of houses according to rank or precedence appears to obtain, and the house of the chief may be either in the centre of the row or at the end. Each house generally accommodates several families, in our sense of the term; which are related together, and under the acknowledged guidance of the elder to whom the house is reputed to belong, and who is really a minor chief, of greater or less importance in the tribe – or village – according to the amount of his property and number of his people.

In front of one or more of the principal houses platforms are often found, on which a group of people may be seen squatting in conversation or engaged in their interminable gambling game. The forest of carved posts in front of the village, each of them representing a great expenditure of property and exertion, doubtless presents to the native eye a grand and awe-inspiring appearance and brings to the mind a sense of probably mysterious import, which possibly does not in reality exist. Behind the dwelling houses, or toward one end of the village and not far removed from it, are the small houses or sheds in which the dead are placed, or pairs of posts supporting a hollowed beam which contains the body.

These permanent villages of the Haidas are now much reduced in number, in correspondence with the very rapid decrease of the people themselves. Those villages least favourably situated as fishing stations, or most remote from communication, have been abandoned, and their people absorbed in others. This has happened especially on the tempestuous west coast of the islands, where there is now but a single inhabited village. Even those still occupied are rapidly falling to decay; the older people gradually dying off, the younger resorting more and more to Victoria and beginning to despise the old ways. Many houses have been completely deserted, while others are shut up and mouldering away under the weather, and yet others, large and fitted to accommodate several families, are occupied by two or three people only. The carved posts, though one may still occasionally be erected, are as a rule more or less advanced towards decay. A rank growth of weeds in some cases presses close up among the inhabited houses, the traffic not being sufficient enough to keep them down. In a few years little of the original aspect of these villages will remain, though at the present moment all their peculiarities can be easily distinguished, and a very little imagination suffices to picture them to the mind as they must have been when swarming with inhabitants dressed in sea-otter robes and seal skins.

The Haidas reside in these permanent villages during the winter season, returning to them after the close of the salmon fishery, about Christmas-time. A portion of the tribe is, however, almost always to be found at the permanent village, and from time to time during other seasons of the year almost the whole tribe may be concentrated there. The villages differ somewhat in this respect. When the territory owned by its people is not very extensive, or does not lie far off, they live almost continually in the village. When it is otherwise, they become widely scattered at several seasons.

The Haidas trouble themselves little about the interior country, but the coast line, and especially the various rivers and streams, are divided among the different families. These tracts are considered as strictly personal property, and are hereditary rights or possessions, descending from one generation to another according to the rule of succession elsewhere stated. They may be bartered or given away, and should one family desire to fish or gather berries in the domain of another, the privilege must be paid for. So strict are these ideas of proprietary right in the soil, that on some parts of the coast sticks may be seen set up to define the limits of the various properties, and woe to the dishonest Indian who appropriates anything of value – as for instance a stranded shark, or seal or sea-otter which has died from its wounds – that comes ashore on the stretch of coast belonging to another. Along the shores the principal berry-gathering grounds are found, and thus divided. The larger salmon streams are often the property jointly of a number of families; and at these autumn fishing grounds temporary houses, small and roughly constructed, are generally to be found. The split cedar planks of the permanent houses are not usually carried by the Haidas to these less substantial houses, though this custom prevails elsewhere on the coast. The construction of the houses thus temporarily occupied is generally so slight and rough as to necessitate no particular description. Poles or cedar planks are built or piled together in whatever manner seems best suited to keep out the rain. In some cases where they are more substantial they resemble on a reduced scale those of the permanent villages. The mode of construction of the latter is described further on. In these temporary shelters, or in even less commodious camps among the trees, the natives live during a considerable part of the year, engaged in salmon fishing, the cutting down of trees and rough hewing of canoes, the gathering and preparation of cedar bark for mats, and other occupations, which, each at its appropriate season, fill out the annual round of duties.

The actual construction of the permanent houses devolves entirely on the men, but is not effected by individual effort. Indeed, the very size of the beams and planks used necessitates the coöperation of many hands. The erection of a house, therefore, in all its stages, from the cutting and hewing out of the beams in the forest, the launching of these and towing them to the village, their erection and fitting, forms the occasion of a 'bee' or gathering of natives, which generally includes detachments from neighbouring villages, and is the occasion of a potlatch or giving away of property by the person for whom the labour is undertaken. Several such gatherings are usually required for the completion of a house, which may be some years in course of construction, as the man for whom the work is done generally exhausts his available resources on each occasion, and requires again to accumulate property, and especially blankets, for a new effort. Dancing and gaming relieve the monotony of the work, which generally occupies but a small portion of each day, and is conducted with much talk and noise, and the shouting of many diverse orders as the great beams are handled.

Among the Haidas each permanent village constitutes a chieftaincy, and has a recognized head chief. The chiefs still possess considerable influence, but it is becoming less, and was doubtless very much greater in former times. It was never, however, the absolute and despotic authority which is sometimes attributed to Indian chiefs. The chief is merely the head or president of the various family combinations, and unless his decisions carry with them the assent of the other leaders, they have not much weight. He has no power of compelling work from other members of the tribe. Should he require a new house he must pay for its erection by making a distribution of property, just as any other man of the tribe would do; and indeed it is expected of the chief that he shall be particularly liberal in these givings away, as well as in providing feasts for the people. He is also supposed to do the honours to distinguished visitors. In Captain Dixon's narrative, the following statements concerning the position of the chiefs at the time of his visit are found: – 'Though every tribe met with at these islands is governed by its respective chief, yet they are divided into families, each of which appears to have regulations and a kind of subordinate government of its own: the chief usually trades for the whole tribe; but I have sometimes observed that when this method of barter has been disapproved of, each separate family has claimed a right to dispose of their own furs, and the chief always complied with this request.'

The chieftaincy is hereditary, and on the death of a chief devolves upon his next eldest brother, or should he have no brother, upon his nephew, or lacking both of these his sister or niece may in rare cases inherit the chieftaincy, though when this occurs it is probably only nominal. It is possible – as occasionally happens in the matter of succession to property – that a distant male relative may, in want of near kinsmen, be adopted by the mother of the deceased as a new son, and may inherit the chieftaincy. I have not, however, heard of cases of this kind. Should all these means of filling the succession fail, a new chief is then either elevated by the consensus of public opinion, or the most opulent and ambitious native attains the position by making a potlatch, or giving away of property greater than any of the rest can afford. Should one man distribute ten blankets, the next may dispose of twenty, the first tries to cap this by a second distribution, and so on till the means of all but one have been exhausted. This form may in reality become a species of election, for should there be a strong feeling in favour of any particular man, his friends may secretly reinforce his means till he carries his point. In no case, however, does the chieftaincy pass from the royal clan to any of the lesser men of the tribe. On being elevated to the chieftaincy the chief assumes a hereditary name, which is also colloquially used as that of the tribe he rules. Thus there is always a Cumshewa, Skedan, Skidegate, &c.; and since the islands have been frequented by vessels, the word 'captain' is frequently added to the titular name of the chief in speaking of him to the whites, to signify his rank.

Certain secrets are reputed to appertain to the office of chief, among which is the possession of various articles of property which are supposed to be mysterious and unknown to the rest of the Indians, or common people (Haida *a-li-kwa*). A very intelligent Skidegate Indian from whom I derived much information, as he was well versed in the Chinook jargon, told me, for instance, that on the death of the last Skidegate chief, the new chief wished him to perform a dance in honour of the great departed, this being one of the rites which it is necessary that the heir should attend to. The dance is one made by a single man, the performer being naked with the exception of the breech-cloth. When my informant was about to engage in the dance the chief took him aside, showing him various articles of the mysterious *chief's properties*. Among others a peculiar whistle, or cell with vibrating reed tongues, which concealed in the mouth enables the operator to produce strange and startling noises, that may be supposed by those not in the secret to indicate a species of possession in the excited dancer. These things are explained by the chief to his probable successor, and are also known to some of the more important Indians; but not to all. They are, no doubt, among the devices for obtaining and holding authority over the credulous vulgar.

Among the Tshimsians in former days, and probably also among the Haidas, a chief had always his principal man, who has considerable authority, and gives advice and instruction to the chief's successor. He never inherits the chieftaincy, however. Each chief with the Tshimsians had also his 'jester,' who is sent on errands of invitation, announces the guests on their arrival, and makes jokes and endeavours to amuse the company, though preserving his own gravity. The jester is not, of course, always in attendance. He receives nothing for his trouble, apparently looking on the position as honourable, and inherits nothing on the chief's death.

It not infrequently happens that a chief grown old, decrepit or poor, though the honourable title still clings to him, is virtually succeeded by some more energetic man, who sways the actions of the tribe in his stead. The village appears to be the largest unit in the Haida system of government, and there has not been any permanent premier chief, or larger confederacy or league of tribes. Such unions may doubtless have been formed from time to time for offensive and defensive purposes, but have not endured.

No laws appear to be acknowledged, but any action tending to the injury of another in person or property lays the offender open to reprisals by the sufferer, but may be atoned for, and the feud closed by payment in blankets or other valuable property to a satisfactory amount. The culprit generally prefers this mode of settlement to having an uncertain retribution hanging over him, and as the value set on property is great, and the disinclination to reduce the store of blankets – which may possibly be accumulating for a prospective distribution – excessive, the restraint is proportionately severe.

RELIGION AND 'MEDICINE'

It is difficult to decide precisely how much should be included under the heading *religion*. The older Indians, and indeed those of every age where they have come not too closely in contact with the whites, show a persistent – one might almost say a fervent – reverence for their time-honoured customs, among which, in this case, the giving away of property or *potlatch* and the various dances, are the most prominent. There are no priests, however, nor could I hear of any religious ritual among the Haidas. The medicine or mystery man, or shaman (Haida *ska-ga*), occupies a position perhaps partly partaking of the priestly function, but more closely allied to that of the prophet, sorcerer, or physician. The Tshimsians say that the Haidas had originally no religion whatever, but adopted their ceremonies not a very great while ago. This may account for the use of Tshimsian words in the dances among the Haidas, and the high esteem in which the Tshimsian language is held by them. It is possible that some of the dances described farther on may have, in part, a religious significance and form a portion of the religious ceremonies above referred to.

It is, however, unquestionable that the Haidas have, and had before any missionary leaven spread among them, an idea of a chief deity, or lord of all things, whose dwelling was in some remote undefined region. This I ascertained by careful inquiry from the Skidegate Indian already referred to, and Mr. Collison, who has been two years among the Masset Haidas as a missionary, and can speak the language with some fluency, confirms me in this statement. The name of this being is *Sun-ī-a-tlai-dus*, or *Sha-nung-ī-tlag-i-das*. His attributes are generally good, but it is difficult to ascertain exactly what they are, owing to the reticence observed by natives in speaking to whites of those of their customs or beliefs which they fear may be ridiculed, but perhaps also in this case to the fact that they have at no time been very precisely defined. The idea of a spirit, soul, or essence being in reality the man, and distinctly separable from the more perishable body, is also firmly rooted in the Haida mind. There is also a recognised principle of evil, called *Hai-de-lān-a*, a name signifying chief of the lower regions. This being is either typified by, or assumes the form of a certain inhabitant of the sea, believed to be the killer whale (*Orca ater*). Indians who lose their lives by drowning are taken possession of by the power of evil, and are turned into beings like himself under his chieftainship. Those killed in battle, or even non-combatants accidentally killed during a fight, go at once to the country of *Sun-ī-a-tlai-dus*, which is supposed to be a happy region. The spirits of those who die from disease, or in the course of nature, become latent, or pass to an ill-defined Hades, but are from time to time recovered, returning to the world as the souls of new-born children, generally – or always – in the tribe to which they themselves formerly belonged. This new birth may occur in each case five successive times, but after this the soul is annihilated, 'like earth, knowing nothing.' So at least say some of the Haidas. The medicine-men profess, in many cases, to be able by means of dreams or visions to tell in the person of what child such an one formerly dead has returned – hence a considerable part of the influence they exercise.

The Indian informant, already several times referred to, told me that the medicine-man had assured him that his brother had returned in the form of a child lately born. He was in doubt whether to believe implicitly or not. I have been told also of a case at Masset, where an old chief dying said to those about him that he would return in the form of a child then about to be born from the wife of one of his relatives. He enjoined them to be careful of the child.

It would seem also to be believed that before death the soul loosens itself from the body, and finally takes its departure altogether. This, at least, would appear to be implied by the fact that the medicine-men sometimes profess to catch the soul of one about to die. This, however, belongs more strictly to the curative function of the *skā-ga*.

The office of *skā-ga*, shaman or medicine-man is not, like the chieftaincy, hereditary, but is either chosen or accepted in consequence of some tendency to dream or see visions, or owing to some omen. The would-be doctor must go through a severe course of initiation. He must abstain from connexion with women, and eat very little ordinary food, and that only once a day, in the evening. He goes into the woods and eats 'medicine,' of which the *Moneses uniflora* was pointed out to me as one of the chief constituents. This plant is hot and bitter to the taste. A course of this character continued for some months, or for even a year, causes the body to become thin, and the mind may eventually be somewhat deranged, or at least the *skā-ga* pretends to see strange things. He speaks mysteriously, and soon takes an acknowledged place in the tribe. When sickness occurs he must be in attendance on the patient, and seeks by every means to exorcise the evil spirit which, abiding in the body, may have caused the disease. The greatest effort is to drive out this spirit, and for this purpose he comes armed with his rattle, or with a drum. The house where the patient lies is probably filled with his friends, the *skā-ga*, drumming or rattling and singing about him, seems to strain every nerve to drive away the evil one. The relatives encourage him to redoubled exertions by promises of property, which, in event of recovery, he will be given.

A *skā-ga* has his hair long and tangled, as, in obedience to custom, it is neither allowed to be cut or comb passed through it. This constitutes a part of his 'medicine.' Besides the rattle or drum the most important property of a *skā-ga* appears to be a hollow bone, carved externally; in some cases also inlaid with pieces of haliotis shell, and open at the ends. In this, using a little shred cedar bark to plug the ends, he can enclose the soul or *ka-tlun-dai* about to depart, and may succeed in restoring it to the body.

From their position the medicine-men are often able to levy blackmail on the credulous, and profit by this species of priestcraft. At Metlakatla the following incident occurred, and was related to me by Mr.

Duncan. This was among the Tshimsians, whose customs in regard to these matters are, however, closely like those of the Haidas: – A medicine-man from an outlying district, coming among the Indians at the mission, put a family into great distress by communicating to them that in walking along, not far off, he had seen the soul of a young girl, had caught it, and for a certain consideration would restore it to the owner, who must otherwise assuredly soon die. The girl indicated was in good health, but some of the relatives were so much alarmed that they came to Mr. Duncan, telling him all the circumstances. He partially reassured them, and finally quieted their fears by frightening the medicine-man himself away.

The *skā-ga* dying, remains still an object of superstition, and his body is not disposed of in exactly the same way with those of mere ordinary mortals. He is not, as they are, boxed up and deposited in little houses in the immediate vicinity of the village, but removed to some distance, in some instances to a place designated by himself before death. The method of sepulture may not be quite uniform, but I can describe that of a medicine-man considered very potent, who died about ten years ago at Skidegate: – On a small island, some miles from the village, is a little box-like hovel, about five feet in height, and nearly square, made of split cedar boards, neatly joined, and roofed with similar planks, on which large stones had been piled to keep the whole firm. The erection stands under a few scattered pine trees, near the rocky shore. A board having fallen out, a good view could be gained of the interior. The side furthest from the water was entirely covered by a neatly made cedar-bark mat. The body leaned against this, in a sitting posture, the knees had originally been drawn up nearly to the chin, but the whole had slipped down somewhat during decomposition. It was not enclosed in any box, but a large red blanket, wrapped round the shoulders, covered the entire lower portion of the body to the ground. The hair, which was long, was still in place, black and glossy, carefully wound up to form a large knot on the top of the head, through which a couple of carved bone pins or skewers were stuck. A carved stick, like those used in dancing, rested in one corner, and before the knees was a square cedar box, which no doubt contained various other properties. Had I not had with me an Indian of the tribe, I should have been tempted to investigate further. The face was the only part of the body uncovered, and the flesh appeared to have been partly dried on the bones, giving it a mummy-like aspect. I mention this fact as it is believed both at Skidegate and Masset, and probably generally among the Haidas, that the bodies of medicine-men do not decay like those of others, leaving only the bones, but dry up without decomposition. In this particular case, it is said among the people of the tribe that if anyone looking at the dead man should see a skeleton only, he or some of his near kinsfolk will surely soon die, whereas if flesh is seen the omen is propitious.

Of another *skā-ga* entombed near the Skidegate Village, I was told by a Haida that on one occasion he was returning to the village, about twilight, when, on looking to where he knew the tomb to be, he saw the *skā-ga* himself, standing erect with his medicine rattle in his hand. My informant was much frightened, and on getting to the village told the people what he had seen, causing no small commotion among them, for the apparition was universally accepted as an evil omen. Shortly afterwards his wife, brother, brother's wife, and two sisters went, with others, to Victoria, and all taking small-pox died there.

A medicine-man is entitled to take from the grave of his predecessor any of his peculiar properties. The privilege is, however, not always or immediately made use of, and it may probably be necessary to wait for some dream or omen before doing so.

The following method of procedure to obtain a fair wind, though not confined in practice to medicine-men, but known to most of the Haidas, may serve to show the childish nature of their mystery performances. An Indian fasting, shoots a raven, quickly singes it in the fire, and then going to the edge of the sea, sweeps it four times on the surface in the direction in which the wind is desired. He then throws it behind him, but afterwards picking it up, sets it in a sitting posture at the foot of a spruce tree, facing toward the required wind. Propping its beak open with a stick, he then requests a fair wind for a certain number of days, and going away lies down and covers himself up with his blanket, till a second Indian asks him for how many days he has required the wind, to which question he answers.

There are among the neighbouring Tshimsians four 'religions,' or systems of rites of a religious character. These have no relation to the totems, but divide the tribe on different lines. They are known as (1) *Sim-ha-lait*, (2) *Mi-hla* (3) *Noo-hlem*, (4) *Hop-pop*. The first is the simplest and seems to have no very distinctive rites. The central figure of the worship of the second was at Fort Simpson a little black image with long hair known as 'the only one above.' The third are 'dog-eaters,' a portion of their rite consisting in killing and cutting, or tearing to pieces, dogs, and eating the flesh. They eat in reality, however, as little of the flesh as they can, quietly disposing of the bulk of it when out of sight. The *hop-pop* or 'cannibals' are those who, in a state of real or pretended frenzy, bite flesh out of the extended arms of the people of the village as a part of their rite. When they issue forth for this purpose they utter cries like hop-pop – whence their name. On this sound being heard all but those of the same religion get out of the way if they can, frequently

12 'Masset Indian village,' 10 August 1878. Few Masset Haida were in the village when Dawson took this photograph. He would meet them later in Virago Sound where they had gone for a house dedication and potlatch.

13 'Mission buildings at Masset,' 18 August 1878, where Dawson was entertained by the Reverend William Henry Collison at tea and dinner and where he attended his only church service of the trip. The Anglican Church Missionary Society's mission buildings had been under the charge of Collison since 1876. Conversion of the Haida was under way by 1878, although this endeavour would not be successful for another three decades.

14 'Looking up Masset Inlet from Indian village,' 18 August 1878. Dawson's photograph shows the funnel-shaped entrance to the broad expanse of Masset Sound, from which he had just returned after seven days of exploration.

15 'Kung Indian Village, Virago Sound,' 21 August 1878. Chief Albert Edward Edenshaw had recently moved here with his people from Kiusta, but, dissatisfied with the site, they were already in the process of moving to Yatza village. A small place, in 1878, the Kung Indian village had eight or ten houses and only a few posts, though the central one in the photograph, almost undecorated, is quite new.

16 'Group of Haida Indians, including Chief Edenshaw, Second from Left, Standing,' 23 August 1878. These were some of the Haida gathered at Yatza for the raising of the first pole in that newly established village. Dawson sought to get as many of the people as possible into the picture, but most 'disliked the idea, & especially the women, not one of whom appeared.'

17 'Edenshaw and Hoo-ya, Chiefs of Yats at Masset,' 23 August 1878. Chiefs Albert Edward Edenshaw of Virago Sound and Hoo-ya (better known as Weah) of Masset, the former, according to Dawson, 'a decent looking & well dressed old man,' though suspected of complicity in the plunder of the *Susan Sturgis* in 1852, and the latter 'a short Indian, remarkable from his grey hair & beard.'

18 'Pillar Rock, Pillar Bay, Graham Island,' 24 August 1878. This very remarkable rock was formed of coarse conglomerate and was estimated by Dawson to stand over eighty feet high.

19 'Dadans Indian village, North Island,' 24 August 1878. Dadans had been important in the early fur trade but was now abandoned except, as in Dawson's photograph, when it was used seasonally as a fishing camp.

20 'Port Simpson village and church,' 30 August 1878. The site of the Hudson's Bay Company's Fort Simpson, the village was now dominated by its large white Methodist church, 'which is quite a landmark from far off the harbour.' It had been built under the supervision of the Reverend Thomas Crosby, who had been invited there by the Tsimshian in 1874. Many visitors regarded the village as a model of Christian industry.

21 'Metlakatla church and mission buildings,' 2 September 1878. Dawson's photograph of this famous mission village shows the cannery building, the church, part of the school, and a few of the individual dwellings. Established in 1862 by William Duncan, lay missionary with the Anglican Church Missionary Society, it became a thriving centre renowned for its achievements in 'advancing' the Tsimshian who had come to live there.

22 'Looking north up Klemtu Passage,' 9 September 1878. With a bright sun and perfect calm, Dawson went ashore on a small island to take a photograph northward up the passage. 'If it developes should show curios effect of kelps and reflexion of trees in water. Seems, however, rather an act of faith to expose an "Extra sensitive" plate ten or fifteen seconds & expect to carry away a picture!' His faith was rewarded in this, the most poetic of Dawson's photographs.

23 'Glaciated rocks and Indian graves, Shell Island, Fort Rupert,' 2 October 1878. This photograph combines Dawson's geological concern for evidence of glacially striated rocks and his interest in Natives. Grave houses were but one way in which the Kwakiutl honoured their dead; coffin boxes in trees and burial caves were also used.

24 'Indian village at Hudson's Bay Company post, Fort Rupert,' 2 October 1878. Fort Rupert had been established as a Hudson's Bay Company post in 1849 at the site of a coal seam. It drew several groups of Kwakiutl people to it and became, until overshadowed by Alert Bay, the largest of Kwakiutl settlements. However, the poles here were fewer and less complex than those of the Haida.

25 'Comox Wharf,' 12 October 1878. Dawson's *Wanderer* stopped at the Comox wharf only long enough for him to check for mail and to secure recent newspapers – and to take a photograph of the modest town.

pushing off in canoes for this purpose. Those of the same creed, and brave, resolutely extend their arms to be bitten. A man may belong to more than one religion, and is in some cases even forced to become initiated into a second. If, for instance, one should pass where dog-eaters are holding a solemn conclave, he may be seized and initiated as a dog-eater *nolens volens*. Great hardships are sometimes endured during initiation. The more savage religions pretend to mysterious supernatural powers, and go to great pains sometimes to delude the common people, or those of other creeds. At Fort Simpson, for instance, a young chief was on one occasion carefully buried in the ground beforehand. When discovered the operators were pulling at a rope, and were supposed to be drawing the chief underground from the back of an island some way off. The rope after a time breaking, great apparent excitement occurs among the operators, who say the chief is now lost, but catching the sticks begin to dig in the ground, and soon unearth him to the great amazement of the vulgar. In this case, however, the cold and cramped attitude so affected the chief that he was lame for life. They instil the truth of such stories especially in the minds of the young, who firmly believe in them. At Fort Simpson, in former days, they have even got up such things as an artificial whale, in some way formed on a canoe. This appeared suddenly on the bay, seemingly swimming along, with a little child on its back.

POTLATCH, OR DISTRIBUTION OF PROPERTY

The distribution of property, or *potlatch* as it is called in the Chinook jargon (Haida, *kie-is-hil*), implying, as it appears at first sight, such entire self-abnegation and disregard of the value of slowly accumulated wealth, requires some explanation. The custom thus named is very widely spread, extending not only to all the coast tribes of British Columbia and its adjacent islands, but also to the native inhabitants of the interior of the Province, of entirely different stocks. I have been able to ascertain more about this custom among the Haidas than elsewhere. Whether in all the other tribes it is so perfectly systematized, or carried out precisely in the same way, it is impossible at present to tell, but among the inhabitants of at least the whole part of the coast the usage appears to vary very little.

The potlatch besides being a means of combining labour for an industrial 'bee,' for purposes in which individual effort is insufficient, is also a method of acquiring influence in the tribe, and in some cases, as we have seen, of attaining even to the chieftaincy. The more frequently and liberally an individual thus distributes property, the more important he becomes in the eyes of his tribe, and the more is owing to him when some other member performs the same ceremony. Only in certain special circumstances are the blankets – which generally constitute the greater part of the property distributed – torn into shreds and destroyed. In most cases it is known long beforehand that a certain man is about to make a distribution, for the purpose of raising a house, cutting out and erecting a new carved post, or other exertion. Some months previously, among the Haidas, he quietly distributes among his friends and the principal members of the tribe his property, be it in blankets or money. The mode of distribution and value of property given to each person is thoroughly systematised, and all the members of the tribe know beforehand how many blankets go to each. A short time before the ceremony all this property is returned with interest; a man who has received four blankets, giving back six, or some larger number in something like this ratio. This retention of a certain amount of the property and its return with increase, appears to be looked upon as an honour by those to whom it is given out. The members of the tribe are then called together for a certain date, and at the same time parties from other, and perhaps distant, villages are invited. The work in hand is accomplished, the man for whom it is done making feasts of the best he has for his guests, and the toil being varied by dancing and gambling with the gaming-sticks, which occupy all the time not more profitably employed. The work finished, the distribution takes place, and shortly afterwards all disperse.

It is usual to make a potlatch on the occasion of tattooing a child, and at other stages in its advance toward manhood. When it is desired to show an utter disregard of worldly wealth, the blankets are torn into strips and scattered among the crowd, and money is also strewn broadcast. This procedure is sometimes followed in competitions for the chieftaincy, already referred to. A similar practice is also a method for showing rage or grief. At Masset, lately, it became known to a father that a young man had made improper advances to his daughter. The father immediately, in great anger, tore up twenty blankets, which not only served as an outlet for his feelings, but placed the young man under the necessity of destroying a similar number of blankets; and in this case, not being possessed of sufficient property, those of the young man's totem-clan had to furnish by subscription the requisite number, or leave upon themselves a lasting disgrace. The feelings of the subscribers were not naturally of the kindest toward the young man, but they did not in this case turn him out of the tribe, as they had a right to do *after* having atoned for his fault.

Among the Tshimsians an ordinary man confines his potlatch or *yak* to those of his own village, while a chief generally, or often, invites people from other villages also. The chief may be assisted in giving potlatches by his people. Should he desire help of this kind, he gives a feast with many different dishes, to which all are invited. The next day a drum is beaten for him by his jester in a peculiar manner, when all who have been at the feast come together with gifts, which are afterwards, with those belonging to the chief himself, given away.

DANCING CEREMONIES

The dance is closely connected with the potlatch ceremonies, but also takes place in some instances without the occasion of a giving away of property. In most of the dances the Tshimsian language is used in the song, which would appear to indicate that the ceremonial has been borrowed from these people. Notwithstanding the old-time hostility of the Haidas and Tshimsians, the former profess a great liking for the Tshimsian language, and many of them speak it fluently.

Six kinds of dancing ceremonies are distinguished, and are designated in the Skidegate dialect by the following names: – (1) *Skā-ga*, (2) *Ska-dul* (3) *Kwai-o-guns-o-lung*, (4) *Ka-ta-ka-gun*, (5) *Ska-rut*, (6) *Hi-atl*. Of these I have only witnessed No. 3, the description of the others being at second-hand from the intelligent Skidegate Indian already more than once referred to.

(1) *Skā-ga* is performed on occasions of joy, as when friendly Indians arrive at a village in their canoes, and it is desired to manifest pleasure. A chief performs this dance. He takes his stand in the house at the side of the central fire furthest from the door. He should wear over his shoulders one of the *na-xin* or Tshimsian blankets, made of fine cedar-bark and the wool of the mountain goat. He wears, besides, the best clothes he may happen to have, and on his head an ornament made of the stout bristles from the whiskers of the sea-lion. These are set upright in a circle, and between them feather-down is heaped, which as he moves is scattered on all sides, filling the air and covering the spectators. He dances in the usual slouching way common among the Indians, bending his knees, but not lifting his feet far from the ground. The people, sitting around in the fire-light, all sing, and the drum is continually beaten. This dance may last half an hour or an hour.

(2) The dance distinguished as *Ska-dul*, appears to be merely the beginning of that known as (3) *Kwai-o-guns-o-lung*. Any man who knows the mode of singing starts the dance alone, when it is called *Ska-dul*, soon others join in, and it becomes No. 3. This is performed by no particular number of people, the more the better, and occurs only when a man desires shortly to make a house. The man himself does not dance, nor does any giving away of property take place. The women occupy a prominent place in this dance, being carefully dressed with the little

marks and *na-xin* or cloaks previously described. One man performs on a drum or tamborine to which all sing, or grunt in time, shuffling about with a jerky motion as they do so. There is a master of the ceremonies who leads off the chorus. Rattles are freely used. The song is in praise of the man who intends to build, and also of the dancers. It eulogises his strength, riches, and so on, and is in the Tshimsian language.

(4) *Ka-ta-ka-gun*. This is performed by the male relatives of a man's wife, and takes place when a house has been finished, the owner at the same time making a distribution of property. The dancers are attired in their best, ornamented, and with faces painted, but no birds'-down is used. It is performed in the newly finished house, and may occupy half an hour or an hour. The man who makes the distribution does not dance. All sing in the Tshimsian language.

(5) *Ska-rut*. One man performs this dance, but is generally or always paid to do the duty for the person more immediately concerned. It takes place some days before a distribution of property, on the occasion of such an event as the tattooing of a child or death of a relative or friend. The dance is performed by a single man, naked with the exception of his breech-cloth. In the first part of the dance, which appears to be intended to simulate a sort of possession or frenzy, one of the grotesque wooden masks is worn, and this is the only dance in which they are used. The wearing of the mask is not, however, absolutely necessary, but is a matter of choice with the performer. Getting heated in the dance, he throws the mask away, snatches up the first dog he can find, kills him, and tearing pieces of his flesh eats them. This dance is not performed in the house as the others are, but at large through the village. The usual present tariff for the performance of the ceremony is about ten blankets. On enquiring what the feelings of the man might be whose dog was devoured, I found that afterwards the dog is appraised and paid for to the satisfaction of all parties. This is characteristic of the manner in which, among the Haidas themselves, the principle of nothing for nothing is strictly carried out.

(6) *Hi-atl*. This dance is very frequently indulged in, and is on occasion of any joyful event, as the arrival of visitors, &c. It is performed by several or many men, who wear feathers in their hair and paint their faces. The Haida language is used in the song. No distribution of property happens, except in the case of the dance being to denote the conclusion of mourning for a dead friend. In this instance a potlatch occurs by the former mourner, who invites his friends together to dance with him.

Gambling is as common with the Haidas as among most other tribes, which means that it is the most popular and constantly practised of all their amusements. The gambler frequently loses his entire property, continuing the play till he has nothing whatever to stake. The game generally played I have not been able to understand clearly. It is the same with that of most of the coast tribes, and not dissimilar from gambling games played by the natives from the Pacific coast to Lake Superior. Sitting on the ground in a circle, in the centre of which a clean cedar mat is spread, each man produces his bundle of neatly smoothed sticks, the values of which are known by the markings upon them. They are shuffled together in soft teased cedar bark, and drawn out by chance.

SOCIAL CUSTOMS

Some points connected with the social relations of the Haidas have already been touched upon, others may be noted here.

A man wishing to marry, informs his mother on what girl his heart is fixed, and she, going to the mother of the beloved one (sweetheart or *ka-ta-dha*), endeavours to arrange the match. An understanding having been arrived at, the man, when ready, invites his friends to accompany him, and going together to the house of the girl's parents, they enter, and sit down around the fire, beside which the girl and her friends also are. The young man's friends then speak in his favour, recommending him to the father of the girl, and praising his good qualities. When the talk is finished, the girl rises, and going to where her would-be husband is, sits down beside him and takes his hand. The ceremony is then complete, and the father of the girl gives various articles of property to her, constituting her dowry. She is led away by her husband, but after a time returns on a visit to her parents, bringing presents, generally of food, from her husband.

Marriage is contracted early. Polygamy is practised, but not extensively; it was formerly more usual, but was always mainly or entirely confined to recognised chiefs. I could hear of but a single instance in which a man yet has two wives. This case is at Skidegate. Three or four wives were not uncommon with a chief in former days, and it was told to me as a tradition by a Haida that a Tshimsian chief at one time had ten wives. As the women do not contribute materially to the support of the family, attending only to the accessory duties of curing and preserving the fish, it is probably difficult for a man to maintain many wives. The women appear to be well treated on the whole, are by no means looked upon as mere servants, and have a voice in most matters in which the men engage. Children are desired, and treated as well as the mode of life and knowledge of the Haida admits. Very few children are now, however, seen about some of the villages, the women resorting to Victoria for purposes of prostitution. Their husbands, be it said to their shame, frequently accompany them, and live on their ill-gotten gains. It is said that in the early days of their contact with the whites, the Haidas were distinguished by good morals. If so, they differed from most of the coast tribes, among whom great laxity has always prevailed. Female chastity is certainly not now prized.

When a girl is about to reach maturity she must attend to various ceremonies, and pass through certain ordeals. It was the custom that she should wear a peculiar cloak or hood at that time for several months, or even half a year. This was made of woven cedar-bark, nearly conical in shape, and reached down below the breast, though open before the face. It was, I believe, called *ky-xe*. The face was painted with the powdered fungus already alluded to, and fasting more or less severe was practised. It was also customary to screen off a corner of the lodge and give the girl a separate fire, and allow her to go out and in by a separate door at the back of the house. This was connected with an idea of ceremonial uncleanness. Did she require to pass out by the front door, it was necessary first to remove all the arms and various other things. In meeting men, the face was to be quickly covered with a corner of the blanket. These or other similar customs were also in vogue among the Tshimsians, whose practices so closely resemble the Haidas in most respects. Among these people great care was taken to teach the girls submission, contentment, and industry. At certain times they were not allowed to lie down to sleep, but if overcome with drowsiness must prop themselves in a sitting posture between boxes. Before drinking, the cup must be turned round four times in the direction of movement of the sun. It was also usual for the mother to save all hairs combed out of the head of the girl, and twist them into cords, which were then tightly tied round the waist and ankles, and left there till they fell to pieces of themselves. This was supposed to give a fine shape to the body. In eating, the girl must always sit down, to prevent a too great corpulence. If orphaned the various ceremonies must be again performed by the girl, even though already all attended to.

Among the Tshimsians peculiar ceremonies exist in connection with the 'bringing out' of young men and women, and it is an occasion of public feasting. In the case of a young woman, the people being all collected, a curtain is raised, and she is seen sitting with her back to the spectators, peculiarly dressed, and surrounded by a circle of upright 'coppers,' if enough can be mustered. She then begins to sing, or, if she does not, an old woman begins to sing near her, and she becoming encouraged joins. The old woman then gradually drops her voice till the novice is singing alone. She then eventually makes a dance before all the people. The songs and dances are practised before the time for the rite arrives. Similar customs probably exist among the Haidas, though I did not learn any details concerning them.

With the Haidas a first-born son may be called by the name of the mother's eldest brother, the second-born after the mother's second brother, or by one of the additional names of the first. Should the mother have no brother, the name of some dead friend is chosen, or in cases where the medicine-man reveals the return of some one formerly dead in the new-born child, the name of the person supposed to be thus returning to the tribe takes precedence of all others. A chief's son is named by its mother after consultation with a medicine-man, whom she pays. He takes a night to think, and mayhap dream, about it. Thereafter he gives the name of a deceased male relative on the mother's side, which is adopted. The ceremony of naming is witnessed by many, and presents are given. A sister of the father's holds the child when named, and becomes its 'godmother' afterwards. For this she receives presents from the father, and from the boy himself when grown up if she has used him well. The next ceremony is that of piercing the lobes of the ears and septum of the nose, when gifts are again distributed, the godmother-aunt coming in for a good share. Four times in all a youth changes his name, always taking one from his mother's family. A potlatch and tattooing of the youth takes place on each occasion except the first, when the latter is omitted. Also a house-building bee. On the last of these occasions the young man is aided by his mother's people, makes the potlatch from his own house and in his own last-adopted name. Dancing and singing are in order at all potlatches. The first house-building is called *tux-kuxo*. The second *ki-au-ni-gexa*. The third *xashl*. The fourth *tlo-xo-kīs-til*.

Slavery is intimately interwoven with the social system of the Haidas, as with that of most of the tribes of the coast. Slaves were formerly common among them, expeditions being undertaken – especially northward to the country about Sitka, where the totems are different – for the special purposes of securing slaves. The intertribal wars along the coast have now ceased, however, and such piratical expeditions have also been abandoned owing to the wholesome dread of gunboats. Slaves, in consequence, are becoming scarce, and the custom is dying away. A slave is called *elaidi* in the Haida language. They appear to have been formerly under the absolute rule of their respective masters, and were sometimes cruelly treated. In some cases a slave has been killed to bury beneath the corner post of a new house. They are veritable hewers of wood and drawers of water. They can be sold, and are supposed at the present time to be worth about two hundred blankets each, the price having risen owing to their scarcity. Children born of slaves are also slaves.

One slave still remains among the Gold Harbour Haidas. There are none at Skidegate or other of the southern villages, but a considerable number at Masset and the northern villages. Slaves sometimes regain their freedom by running away, but should they return to their native place are generally so much despised that their lives are rendered miserable.

When a man falls sick it devolves upon his brother to call in the medicine-man, and also to invite the friends to the house of sickness, and provide them with tobacco to smoke. The house is thus generally full of sympathising Indians, with smoke, and the noise of the medicine-man's performances. Should the sick man die, the body is generally enclosed in a sitting posture in a nearly square cedar box, which is made for the purpose by all the Indians conjointly; or, if they do not wish to make it, they subscribe to purchase from some one of their number a suitable box. The coffin-box being the same in shape as those used for ordinary domestic purposes, there is generally no difficulty in securing one. In either case the brother, or other near relative of the deceased, makes a potlatch, or distribution of property, to repay the others for their labour or expense.

If a man of ordinary reputation only, dies, his body (*tl-kō-da*) is put at once into the coffin-box (*sa-tling-un*), and is then stored away in the tomb-house (*sa-tling-un-nai*), which is generally a little, covered shed behind the house, or in the immediate neighbourhood of the village. This tomb is also made by the combined labour of the men of the village and paid for in the same way as in the case of the coffin-box. In it may be placed but a single body, or two or more – those of relatives. Should the dead have been a man of great importance, or a chief, the box containing the body is placed in the house inhabited during life, the other occupants finding quarters elsewhere as best they can. The clothes and other articles of property of the dead man are arranged about him, and he sits in state thus for perhaps a year, no one removing any of the things. Indians from another village, however, may come to see the body, and do so. The body once consigned to the tomb-house is now left there, but it was formerly the custom in the case of chiefs to open the tomb from time to time and provide the body with fresh blankets or robes. This is said never to have been done to the bodies of the less important members of the tribe, and to have been long in disuse; it is a common practice among the Salish Indians of the interior of British Columbia. Both among the Haidas and Tshimsians the dead were also formerly burnt as an occasional or not unfrequent practice. In this case the ashes were collected and put in a box. This is never now done, but numerous instances occurred in the last generation.

After the body has been entombed it becomes necessary sooner or later, if the deceased has been a person of any importance in the tribe, to erect a carved post. The Indians again collect for this purpose, and are repaid by a distribution of property, made by the brother of the deceased or other relative to whom his estate has come down as next in order of descent. The post erected, though sometimes equally ponderous with the carved posts of the houses, is not generally so elaborate. In many cases it consists of a plain upright, tapering slightly towards the lower end, or that inserted in the ground, while the upper bears a broad board, on which some design is carved or painted, or any 'coppers' formerly belonging to the dead man are attached.

The custom of placing the bodies of the dead in canoes, which may either rest on the ground or be fixed in a tree, does not obtain among the Haidas, nor did I see any instance of the use of trees as receptacles of coffin-boxes, as practised among several other tribes of the coast.

The brother of the deceased inherits his property, or should there be no brother, a nephew, or the sister, or, failing all these, the mother. Occasionally some distant male relative may be adopted as a new son by the mother, and be made heir to the property. The wife may in some cases get a small share. As soon as the body has been enclosed in the coffin-box, and not before, the brother or other heir takes possession. When it can be amicably arranged, he also inherits the wife of the dead man, but should he be already married, the nephew or other relative on whom the succession would next devolve is supposed to marry the relict. Should there be no relative to marry her, she may be married again to any other man.

A single system of totems (Haida, *kwalla*) extends throughout the different tribes of the Haidas, Kaiganes, Tshimsians and neighbouring peoples. The whole community is divided under the different totems, and the obligations attaching to totem are not confined by tribal or national limits. The totems found among these peoples are designated by the *eagle, wolf, crow, black bear* and *fin-whale* (or *killer*). The two last-named are united, so that but four clans are counted in all. The Haida names for these are, in order, *koot, koo-ji, kit-si-naka* and *sxa-nu-xā*. The members of the different totems are generally pretty equally distributed in each tribe. Those of the same totem are all counted as it were of one family, and the chief bearing of the system appears to be on marriage. No one may marry in his or her own totem, whether within or without their own tribe or nation. A person of any particular totem may, however, marry one of any other indifferently. The children follow the totem of the mother, save in some very exceptional cases, when a child newly born may be given to the father's sister to suckle. This is done to strengthen the totem of the father when its number has become reduced. The child is then spoken of as belonging to the aunt, but after it attains a certain age may be returned to the real mother to bring up.

An Indian on arriving at a strange village, where he may apprehend hostility, would look for a house indicated by its carved post as belonging to his totem, and make for it. The master of the house coming out, may if he likes make a dance in honour of his visitor, but in any case protects him from all injury. In the same way, should an Indian be captured as a slave by some warlike expedition, and brought into the village of his captors, it behoves any one of his totem, either man or woman, to present themselves to the captors, and singing a certain sacred song, offer to redeem the captive. Blankets and other property are given for this purpose. Should the slave be given up, the redeemer sends him back to his tribe, and the relatives pay the redeemer for what he has expended. Should the captors refuse to give up the slave for the property offered, it is considered rather disgraceful to them. This at least is the custom pursued in regard to captives included in the same totem system as themselves by the Tshimsians, and it is doubtless identical or very similar among the Haidas, though no special information on this subject was obtained from them.

Tattooing, as already mentioned, is universal among the Haidas, the legs, arms and breasts being generally thus ornamented. Among the Tshimsians it is occasionally practised. The design is in all cases the totem-crest of the bearer.

The strictness of the custom of payment for privileges granted, and repayment for losses or injuries sustained, almost necessitated the definition of a currency of some kind. Among most of the coast tribes the dentalium shell was prized, but not so much as a means of exchange among themselves as for barter with the Indians of the interior. By the Haidas the dentalium is called *kwo-tsing*, but as these people were by their position debarred from the trade with the interior, it was probably never of so great value with them. It is still sometimes worn in ornaments, but has disappeared as a medium of exchange.

Another article of purely conventional value, and serving as money, is the 'copper.' This is a piece of native metal beaten out into a flat sheet, and made to take the form illustrated in the margin. These are not made by the Haidas, – nor indeed is the native metal known to exist in the islands, – but are imported as articles of great worth from the *Chil-kat* country, north of Sitka. Much attention is paid to the size and make of the copper, which should be of uniform but not too great thickness, and give forth a good sound when struck with the hand. At the present time spurious coppers have come into circulation, and though these are easily detected by an expert, the value of the copper has become somewhat reduced, and is often more nominal than real.

Formerly ten slaves were paid for a good copper, as a usual price, now they are valued at from forty to eighty blankets.

The *blanket* is now, however, the recognised currency, not only among the Haidas, but generally along the coast. It takes the place of the beaver-skin currency of the interior of British Columbia and the Northwest Territory. The blankets used in trade are distinguished by points, or marks on the edge, woven into their texture, the best being four-point, the smallest and poorest one-point. The acknowledged unit of value is a single two-and-a-half-point blanket, now worth a little over $1.50. Everything is referred to this unit, even a large four-point blanket is said to be worth so many *blankets*. The Hudson Bay Company, at their posts, and other traders, not infrequently buy in blankets, taking them – when in good condition – from the Indians as money, and selling them out again as required.

Blankets are carefully stowed away in large boxes, neatly folded. A man of property may have several hundred. The practice of amassing wealth in blankets, no doubt had its origin in an earlier one of accumulating the sea-otter and fur-seal robes, which stood in the place of blankets in former days. This may help to explain the rich harvest of these skins which the first traders to the Queen Charlotte Islands gathered.

Besides the payments already mentioned, as exacted from a stranger wishing to fish or gather berries in the territory of another, the Tshimsian Indians, who sometimes resort to the southern end of the islands to hunt the sea-otter, are forced to pay the neighbouring tribe for the privilege, though the chase is carried on on the open sea. Certain men, too, supposed to be specially skilled in various kinds of work, are regularly paid for their services. This is expressly the case with workers in wood and those competent to carve and paint the peculiar posts.

Oolachen grease, bought from the Tshimsians, is paid for in blankets, while a return trade in canoes – in the making of which the Haidas excel – is conducted on the same basis.

While at Cumshewa Inlet, we witnessed the arrival of some Tshimsian Indians who had come in canoes loaded with oolachen grease, hoping to sell it to the Haidas. Veritable merchants, ready if they find no market here, to go on to the next village. The sky was just losing the glow of sunset when the two canoes were seen coming round the point. The Haidas, looking attentively at them, pronounced them Tshimsians, and proved to be correct. The greater number of the occupants of the canoes were women, all fairly well dressed, and wearing clean blankets to make a good appearance on their arrival among strangers. The faces of some of them, covered with a nearly black coat of gum and grease, had a wild aspect, which was rendered rather comical, however, by the various and inappropriate nature of the hats and caps – all of civilized patterns – which they wore. Each of the canoes has a couple of masts, to which the light sails are now tightly clewed up, but from the foremost canoe floats a wide strip of red bunting. The paddles are dipped with a slow, monotonous persistency indicative of the close of a long day's work, and they tell us they have only slept twice since leaving Kit-katla. Arrived at the beach opposite the Haida village, the canoes are stranded, and the villagers crowd round to render assistance. The bark boxes holding the precious grease are carefully set in the water, beside the canoes. Kettles, mats, paddles and all the varied articles of the travelling outfit are carried ashore. The canoes are hauled up by united exertion, the boxes of grease carefully carried beyond high-water mark, and covered with brush; and in half an hour, the travellers, distributed among the houses of the village, are found at their evening meal. Business does not seem to occupy their attention; they will remain here several days to talk about that.

ARTS AND ARCHITECTURE

Under this special heading a few points may be taken up, some of which have already been incidentally referred to in general terms.

The primitive sea-otter or seal skin cloak of the Haidas has already been described in extracts quoted from old authors, together with the dressed skin undershirt (99-100), while of the armour of skin and split sticks little can now be learnt. The *naxin*, or dancing shawls made by the Tshimsians, so much prized, and have been described, and the head-dress worn at the same time with the *naxin* mentioned. This consists essentially of a small, nearly flat mask (one in my possession is 6 inches long by $5^3/_4$ wide), fixed to an erection of cedar bark, feathers, &c., in such a manner as to stand erect above the forehead of the woman. At the back depends a train, which may be made of cloth, but should have ermine skins sewn on it. These masks are frequently well carved to represent a human face not unpleasant in expression, and have the teeth and eyes formed of inlaid Haliotis shell.

On ordinary occasions a head-covering is usually dispensed with, unless it be some old hat of European style. The women, nevertheless, make, and occasionally wear, the peculiar basket-work hats common on the coast. These have the form of a rather obtuse cone, of which the sides are hollowed and the apex truncated. They are generally ornamented by painting in black, blue or red, in the conventional style common among these people. The feet are almost invariably bare.

Leggins ornamented with puffin beaks have been referred to as occasionally adopted as a part of the dancing costume. A species of castinet or rattle is also made from these for use in dancing. Each beak is threaded to a thin strip of sinew, and they are then attached at short intervals to the circumference of each of a couple of thin wooden hoops, the diameter of the larger of which may be 8 or 9 inches; of the smaller a little less. A cross-bar connects the two hoops, and being held in the hand, a slight motion in rotation being imparted by the wrist, causes the dry, horny beaks to rattle together.

Masks are to be found in considerable numbers in all the villages, and though I could hear that they were employed in a single dance only, it is probable that there may be other occasions for their use. The masks may be divided into two classes – the first, those which represent human faces, the second those representing birds. They are carved in wood. Those of the first class are usually large enough amply to cover the face. In some cases they are very neatly made, generally to represent an ordinary Indian type of face without any grotesque idea. The relief of the work is generally a little lower than in nature. Straps of leather, fastened to the sides of the mask, are provided to go round the head of the wearer, or a small loop of cedar-bark string is fixed in the hollow side of the mask, to be grasped by the teeth. The top of the forehead is usually fringed with down, hair or feathers. The eyes are pierced to enable the wearer to look out, and the mouth is also often cut through, though sometimes solid, and representing teeth. Grotesque masks are also made in this style, but none were observed to have a smiling or humourous expression. The painting of the masks is, according to taste, in bars and lines, or the peculiar curved lines with eye-like ovals found so frequently in the designs of the coast Indians. The painting of the two sides of the face is rarely symmetrical, a circumstance not arising from any want of skill, but intentionally brought about. Of the second class of masks, representing birds, there are various kinds. One obtained at the Klue Village had a beak five or six feet long projecting from the centre of a mask not much unlike those above described. The beak was painted red, and the whole evidently intended to represent the oyster-catcher common on the coast. Another mask represents the head of a puffin, and is very well modelled. It is too small within, however, to allow the head to enter, and must have been worn fixed to the top of the head.

Rattles are also used chiefly in dancing. These are of two principal types. First and most usual are plain spheroidal or oval rattles, generally considerably flattened in shape. They are carved in wood with great neatness, the wood being sometimes reduced to a uniform and very small thickness throughout. Each is made in two pieces, which are fixed together generally by small threads of sinew passed through holes in their edges. Small round pebbles from the beach are placed within. The representation of a human face, which may be plain or coloured, according to the maker's taste, is generally found on each side of these rattles, though some are almost entirely plain. The second species of rattle is much more elaborate in form, is highly prized, and apparently used only by persons of some distinction. These are made in the form of a bird, the handle being in a position corresponding with the bird's tail. Accessory carving of a very elaborate character is sometimes found on these rattles, which can scarcely be described at length here. They are generally carefully painted with red, blue and other colours. Rattles in other forms are also found; one was seen to resemble a killer whale, with a greatly exaggerated back fin.

A carved stick is sometimes held in the hand in dancing, and struck upon the floor in time with the motion of the feet. Several of those which I have seen are about five feet in length, and are carved much in the style of the posts which are set up in front of the houses. Figures of men and conventionalized representations of animals appear to be seated one above another up the length of the stick.

A small apparatus held in the mouth to produce a peculiar noise when dancing, has been mentioned in connection with that custom on a former page. One which I obtained consisted of a wooden tube roughly oval in section, three-quarters of an inch in greatest width, with a length of an inch and a quarter. This is composed of two pieces tied together with a strip of bark, and within it are placed two vibrating pieces, each composed of two flat pieces of wood or reed tied together. In a box in one of the old houses in Parry Passage several such cells were found fitted in trumpet-shaped tubes about a foot in length made of cedar wood, each being composed of two pieces.

In describing the performance of the medicine-men (116) a peculiar charm, or implement by which the departing soul may be caught and perhaps replaced, was referred to. This is made from a piece of bone, which from its size and general shape might be part of a human femur, but may possibly be that of a bear. This bone is pared down so as to have an almost perfectly symmetrical form, the ends being somewhat more expanded than the middle. A human face, often grotesque, ornaments the centre of one side, the remainder of a human figure being sometimes carved so as to extend round over the back in a more or less cramped attitude. The ends are slit, the slit in each instance passing through both sides of the bone, and representing the mouth of a creature the eyes and nostrils of which are rudely indicated in a conventional manner above. The upper side of the bone is pierced by a couple of holes for its suspension over the breast by a string which passes round the neck. A few small holes, probably for the attachment of tassels or other little ornaments are sometimes made in the lower side. Some examples are neatly inlaid with fragments of haliotis shell. The dimensions of two good specimens are, No. 1 – Length 6⅝ inches; vertical diameter in centre, 1 inch, horizontal diameter, ⅞ inch; vertical diameter at ends, 1¼ inch; horizontal diameter at one end, 1 inch, at the other, ⅞ inch; depth of slit at ends, 1½; inches. No. 2 – the dimensions in the same order, 7½; 1; ¾; 1⅜; 1; ¾; 1½ inches.

Bone pins, more or less carefully carved, are used by the medicine-men to secure the knot into which they tie up their hair; and pieces of bone carved to represent whales, birds, human figures, or combinations of these are not unfrequently found, though now seldom worn. They served formerly for ornaments, some of the smaller being probably ear-rings.

A peculiar and very ingenious speaking doll was obtained at Skidegate. This did not seem to be a mere toy, but was looked upon as a thing of worth, and had previously been used, in all probability, as an impressive mystery. It consisted of a small wooden head, 3½ inches high by 2½ inches wide and 2 inches deep from back to front, composed of two pieces of wood hollowed till quite thin, and the front one carved to represent a grotesque face, with a large round open mouth with projecting lips. The two wooden pieces had then been neatly joined, a narrow slit only remaining within the neck, and serving for the passage of air, which then impinging on a sharp edge at the back of the cavity representing the mouth, makes a hollow whistling sound. To the neck is tied the orifice of a bladder, which is filled with some loose elastic substance, probably coarse grass or bark. On squeezing the bladder sharply in the hand a note is produced, and on relaxing the pressure the air runs back silently, enabling the sound to be made as frequently as desired.

Most of the ordinary household utensils are made of wood, or rather it may be said were so made, for at the present day tin and cheap earthenware dishes are rapidly superseding those of native manufacture. Several distinct types of wooden dishes may be distinguished, and these appear to have been followed by the maker with little variation except in the detail of ornamentation. One form, used to hold berries and other food, is a tray of oblong outline, the length being about one and one-third times the width, and the depth comparatively small. These are cut out of solid wood, the edge being slightly undercut within, and the bottom within rounded though externally angular. The outer ends are generally the sides occasionally ornamented by incised carving or painting. The edge is frequently, in the better examples, set with a row of the strong, calcareous opercula of *Pachypoma gibberosum*. These trays are often ten feet or more in length. Another very favourite form may be said to be boat-shaped, the hollow of the dish being oval in outline, but provided at the ends with prow-like wooden projections which serve as handles. One of these is generally carved to represent the head of an animal, the other the tail and hind legs. These dishes are seldom more than eight or ten inches in length, and curve upwards from the middle towards the ends. Another form is oblong in outline, but nearly as deep as wide. Seldom more than about fifteen inches in length. The bottom in the larger of these vessels is frequently a separate flat piece of wood neatly joined. One end of many of these dishes is carved to represent the head of a beaver or other animal, while the other carries a representation of the legs and tail. Other carvings may ornament the sides. This form is sometimes varied in the smaller sizes by making the vertical profile of the longer edges correspond to a graceful curve instead of keeping to one plane. Another modification of this type is found in a dish to one end of which a broad, flat expansion carved to represent the tail of a bird is fixed, while the head projects from the opposite end. The bird is represented as lying on its back when the dish is in its proper position, the hollow being made apparently in the bird's breast. Very large dishes are still occasionally, and were formerly frequently made for use, in feasts given by chiefs, &c. One of these had a general form like that of the first described kind of dish, but was nearly square, the sides being 3 feet 8 inches. It was composed of four side pieces and a bottom piece neatly pegged together, while the edge was surrounded by a double row of opercula. Another form seen in one of the old houses on Parry Passage is a parallel-sided trough six or eight feet long, with a head carved at one end, a tail and pair of swimming feet at the other, the whole being supposed to represent a sea-lion. Still another pattern was found in a shallow, gracefully shaped tray 5 feet 6 inches long, and about one-third as wide. The ends of this were obtusely pointed and overhung, while above, a flat space between each extremity and the end of the hollow within, bore a complicated pattern in incised lines.

The stone mortars already mentioned as having been employed in the preparation of the native tobacco, now seem to be little if at all used for any purpose. They are generally circular in outline and without ornamentation, being in some cases very roughly made. Other examples are ornamented by carving. A plain circular mortar of rather greater size than usual was found to have a width of $9^1/_2$, a heighth of $6^1/_2$, and an internal depth of $4^1/_2$ inches. A second, carved externally to represent a frog had, disregarding the projecting points of the carving, the following dimensions, in the same order as above, $-6^1/_2$; $5^1/_2$; $3^1/_2$ inches. One mortar of an oval form, with projecting carved ends, was seen. It represents a frog or some large-mouthed kind of fish like a cottus, but the design is complicated by the introduction of a human face near what should the hinder end of the animal. The extreme length of this mortar is $16^1/_2$ inches, the width at the middle 8 inches tapering a little from the head to the tail, and the height at the middle, which is slightly lower than the ends, 5½ inches. The dimensions of the interior hollow of this mortar are 8 by 5¾, an 3¼ inches deep. Another stone utensil obtained at Skidegate is a dish for preparing paint. This is 6 inches long by 2½ wide, in external dimensions, with a trough-shaped bowl 4½ by 1¾ inches, in which the paint has evidently been ground by rubbing from end to end with a second stone. When laid with the hollow side downward, the exterior is found to be carved to represent some animal, probably a frog, in a constrained squatting attitude.

Shells, especially those of the large mussel are frequently used as spoons and small dishes. A very handsome dish, with an oval outline, is also made from part of the larger end of the horn of the mountain sheep. This is probably softened by steaming, and forced into a symmetrical shape, then pared down thin and carved externally. The mountain sheep horns, with those of the mountain goat, are obtained in barter with the Tshimsians and other Indians of the mainland, neither of the animals occurring in the Queen Charlotte Islands.

Large serviceable ladles are also made from the mountain sheep horns, the lower part of the horn being widened to form an ample bowl, and the upper straightened out to produce the handle. One of these of the larger sort measures from the end of the handle to the point of the bowl, round its convex surface, 2 feet 3½ inches. The bowl itself is 8½ inches long by 6 inches wide, and 2¼ deep. The spoons in ordinary use are six or seven inches long with large flat bowls, made in a single piece from the horn of the mountain goat. The handle may be carved to represent a human or other form. Another kind much prized and cared for, is made by attaching a bowl of the usual form, made from a piece of mountain sheep or goat horn, to the wider extremity of an entire horn of the mountain goat by a couple of rivets. The goat horn, retaining its natural curve, is then elaborately carved with human or other figures, according to the taste of the maker. Such spoons may be about a foot in length.

Knives of all sorts are now in use, but some ingenuity is shown in adapting old blades to new handles, manufacturing knives from files, and so on. A knife used in cutting up fish is made by fixing one edge of a thin square or oblong piece of iron in a cylindrical or flattened piece of wood of slightly greater length. This has thus the form of a small mincing knife.

The boxes in which most of the goods and chattels of the household are packed away are made after a uniform plan. A small one measured 20½ inches high by 15 square. The sides are made of a single wide thin piece of cedar, which is bent three times at a right angle, with very little appearance of breaking at the corners, and pegged together at the fourth angle. The bottom is made of a separate piece of wood. The cover is cut out of a solid slab. It rests by a shoulder on the ledge of the box, and expands slightly upward, so that the upper surface of that of the box above mentioned is nearly 17 inches square. These boxes are generally decorated externally by designs in black and dull red paint, and are carefully corded with cedar-bark rope, which is so arranged as to meet and tie over the top of the cover when desired.

Mats, of an oblong form, and plaited rather than woven, from strips of cedar bark, constitute a great part of the household furniture. They vary much in texture, and may be either of the natural brownish or yellowish colour or diversified by black bands.

One-handed adzes, with the blade fixed at an acute angle to the handle, are very commonly used. The blade is often an old broad file, sharpened at the end. These, no doubt, replace those of stone of a former day. A few of the stone adze-heads are still to be found about the houses, and are very well shaped, and different in form from any I have elsewhere seen. The head somewhat resembles a poll pick in shape, being square in section near the front, but oblong towards the head owing to the increasing breadth, the thickness from side to side remaining the same or nearly so. Near the head, one of the smaller sides is curved into one or two saddle-like hollows to receive the properly shaped end of the handle, which was no doubt lashed firmly to the stone with sinew or bark. The lateral surfaces are sometimes grooved from the head downward for one-third or more of the total length. The dimensions of some specimens are as follows: –

No. 1 – Length, 1'1".   Breadth, 2".   Thickness, $1^{7}/_{16}$ inches.
No. 2 –   " $7^{1}/_{4}$".   "   2".   " $1^{5}/_{8}$".
No. 3 –   "   8" (about)   "   2".   " $1^{8}/_{15}$".

The measurements are merely averages, as the sides are not generally strictly parallel, but slope more or less towards the ends. The material of these tools appears to be a matter of indifference, as I have seen them made of hard altered igneous rocks like those so common in the country, of a hard sandy argillite, and of the peculiar greenish jade which the natives of some other parts of the province prize so highly. This latter material is not, according to the Haidas, found in the islands, but has occasionally been obtained in the course of trade.

Large stone hammers are still in use for driving home wedges and similar operations. No stone arrow-heads were found, and it is probable that these people, before they were acquainted with iron, used bone only for this purpose.

Spears and harpoons were doubtless in former times made of bone, like those found in the shell heaps of Vancouver Island. At the present day iron has been substituted. A species of harpoon is used in the chase of the fur seal. It is generally made by the Haidas themselves from an old flat file. The extremity is sharpened to a blade-like point, which is succeeded by a series of barbs on each side, sharply thrown backward. The butt of the file is bored through, and a loop of strong copper wire fixed to it so as to move freely. To this is attached a strong cord of plaited sinew, to the extremity of which a bladder or float is affixed. When in use, the butt end of the iron head is fixed in a socket in the extremity of a long, light cedar pole, but easily detaches itself when it is driven into the animal. The head of the harpoon generally fits into a wooden sheath made of two pieces fixed together with bark lashing.

The head of the salmon spear consists of a sharp blade-like iron tip to the base of which two pointed pieces of horn are lashed, the lashing being thickly covered with spruce gum so as to offer no impediment to the whole entering the fish. The length of the blade, with the horn barbs, is about four inches. Between the pieces of horn fits the sharpened end of a piece of wood, $7^{1}/_{2}$ inches long, which increases gradually in size till at its inner extremity it forms a flat leaf-shaped expansion, which fits into a hollow of similar form in the end of a long light cedar pole. The end of the pole is served with bark to prevent its splitting, and the iron-tipped head is made fast to the intermediate wooden piece, and that to the end of the pole by strong strings. When plunged in the fish, the loose wooden piece no doubt first comes out from the end of the pole, and with a slight increase of strain it comes away from the barbed head, which thus practically remains fixed to the end of the pole by a foot or eighteen inches of cord.

The fish hook is made substantially after the pattern general on the west coast, but owing to the want of the yew, it has not the same graceful shape with that of the Ahts and Makah Indians. In its primitive form, among the Haidas, it consists either of a forked branch, of suitable size, or of two pieces of wood lashed together so as to make an acute angle with each other. To the upper piece, about the middle, is fixed the string for the suspension of the whole, to the free or outer end of the lower piece a pointed bone is lashed so as to project obliquely backward, reaching to within a short distance of the upper piece. The bone is now, however, generally replaced by an iron point, and in some cases the whole hook is fashioned out of a piece of thin iron rod, bent round and sharpened. This hook is more particularly used in halibut fishing. A large sized one in wood measures 10 inches in length, with a distance of five and a half inches between the divergent ends of the two pieces of which it is made. When in use, a carved wooden float is fixed about a foot from the hook, and a short distance further up the line a large stone sinker. The whole being lowered to the bottom till the stone comes to rest, the small float drifts out with the tide, and keeps the hook below it at a short distance from the bottom. The wooden parts of the hooks and the floats are sometimes rudely carved. A second form of hook differs slightly from the first, in being formed of a piece of thin iron rod, bent round in a continuous curve of an oval form, but of which the upper side has been somewhat displaced so as to allow the passage of the lip of the fish within the recurved point. These hooks are often made small, and used in catching flounders and such fish.

In the small rivers the salmon are generally caught in fish traps or wiers. A wier of split sticks being fixed completely across the river, cylindrical baskets made of the same material, with an orifice formed of sticks converging inward, serves to entrap the fish; or in other cases, flat frames are placed in such a position that the fish in endeavoring to surmount the wier by leaping falls into them.

The canoes of the Indians of the west coast are similar in type through all the tribes, but differ considerably in detail of shape and size. They are made from the giant cedar (*Thuja gigantea*), the wood of which is light, durable and easily worked, but apt to split parallel to the grain. This constitutes the greatest danger to the Indian canoes in rough weather, especially when they are heavily laden. Among the Haidas two patterns of canoes are found. In the first and most commonly used, the stern projects backwards, sloping slightly upward, and forming a long spur, while it is flattened to an edge below. The bow also curves upward, but has no spur, the cutwater forming a regular curve. These canoes are frequently thirty or thirty-five feet long. The second pattern is that of the larger canoes, intended for longer voyages. In these both bow and stern are provided with a strong spur sloping upward, and generally scarfed to the main body of the canoe. The canoes are often about forty feet long, with a corresponding beam, and were in former days not infrequently constructed to carry forty men besides much baggage. With the exception of the bow and stern pieces, each canoe is made from a single log, which is roughly shaped out where the tree is cut, afterwards floated to a permanent village, and finished at odd hours, during the winter months. The lines of the canoes are very fine, the requisite amount of beam being given to them by steaming with water and hot stones, and the insertion of thwarts. They are smoothed outside and blackened, while inside they generally bear fine and regular tool marks from end to end. The Haidas are great canoe makers, and annually take over a large number of canoes to Port Simpson and the Nasse, which are sold, or exchanged there for oolachen grease or other commodities. The canoe paddles are usually made of cedar or the yellow cypress. Balers for the canoes are generally cut out of wood in the form of a scoop, with handle behind or made from a piece of cedar bark gathered up at the ends in a fan shape, with a stick secured across the top.

Various particulars concerning the manner of the Haidas in living in villages and the houses which they construct have already been given. The houses are placed with their gable-ends to the beach, which constitutes the street, the roof sloping down at a moderate angle on each side, with a projecting oblong 'lantern' or erection in the centre intended for the escape of smoke, and fitted with a movable shutter which may be set against the wind. The houses are oblong or nearly square, and are often from 40 to 50 feet in length of side, and erected to accommodate a great number of people. The older and better built houses are almost invariably partly sunk in the ground. That is to say, the ground has been excavated to a depth of six or eight feet in a square area in the centre of the house, with one or two large steps running round the sides. A small square of bare earth is left in the centre below the smoke-hole, the rest of the floor being generally covered with split cedar planks. The steps which run round the sides are faced and covered above by large hewn slabs of cedar, and serve not only for sleeping and lounging places, but as the depositary of all sorts of boxes and packages of property belonging to the family. Some of the houses stand on the surface of the ground without any excavation. The pattern of the house itself is maintained with little variation in all parts of the islands, and has doubtless been handed down from time immemorial. The first process is to plant firmly in the ground four stout posts of sufficient height at each end. These are called *kwul-skug-it*, and are intended to bear four large beams which run from front to back of the house, and are called *Tsan-skoo-ka-da*. The heads of the posts are hollowed to receive the horizontal beams, which, with the posts, are circular in section. The longitudinal beams do not project beyond the posts which bear them, and in front of them at each end is a frame composed of large flat beams, which support the edge of the roof and the hewn planks of the front of the house. There are generally four flat upright beams, one in front of each of the main upright posts before described. These support a pair of beams which have the same slope with the roof, and are channelled below to receive the upper ends of the hewn boards which close the front of the house. These beams are called *ki-watl-ka*. The two upright beams nearest the centre *ki-stang-o*, the outer *kwul-ki-stung*. The dimensions of the house, of size rather greater than usual, in the Kung Indian Village, Virago Sound, were found to be as follows: – breadth of front of house, 54'6"; depth, from front to back, 47'8"; height of ridge of roof, 16'6"; height of eaves, 10'8"; girth of main vertical posts and horizontal beams, 9'9"; width of outer upright beams, 1'10"; thickness, about 5"; width of upper sloping beams, 2'7"; thickness, 5"; width of carved post in front of house, 3'10".

A second, and not unusual style of house has only a single frame, consisting of four vertical flattened posts at each end, supporting sloping beams. The outer supporting posts are generally morticed out, and the outer ends of the sloping beams passed through them. Stout beams flattened on the lower side, and generally three in number on each side, are then made to rest on the sloping beams, and bear above them the cedar planking of the roof, held in place by stones heaped upon it, or by small beams laid over them above.

In a passage quoted by Mr. J.G. Swan in the Smithsonian Contributions to Knowledge, no. 267, Marchand (1791) describes the houses on North Island in the following terms: – 'The form of these habitations is that of a regular parallelogram, from forty-five to fifty feet in front, by thirty-five in depth. Six, eight, or ten posts, cut and planted in the ground at each front, form the enclosure of a habitation, and are fastened together by planks ten inches in width, by three or four in thickness, which are solidly joined to the posts by tenons and mortices; the enclosures, six or seven feet high, are surmounted by a roof, a little sloped, the summit of which is raised from ten to twelve feet above the ground. These enclosures and the roofing are faced with planks, each of which is about two feet wide. In the middle of the roof is made a large, square opening, which affords, at once, both entrance to the light, and issue to the smoke. There are also a few small windows open on the sides. These houses have two storys, although one only is visible, the second is underground, or rather its upper part or ceiling is even with the surface of the place in which the posts are driven. It consists of a cellar about five feet in depth, dug in the inside of the habitation, at the distance of six feet from the walls throughout the whole of the circumference. The descent to it is by three or four steps made in the platform of earth which is reserved between the foundations of the walls and the cellar; and these steps of earth, well beaten, are cased with planks, which prevent the soil from falling in. Beams laid across, and covered with thick planks, form the upper floor of this subterraneous story, which preserves from moisture the upper story, whose floor is on a level with the ground. This cellar is the winter habitation.'

This description is substantially accurate, and so detailed that it is scarcely likely to be erroneous in regard to the division by a floor of the excavated portion of the interior of the house from that above the level of the ground. I have not seen this arrangement, however, in any of the houses now existing on the islands.

The peculiar carved pillars which have been generally referred to as carved posts are broadly divided into two classes, known as *kexen* and *xat*. One of the former stands at the front of every house, and through the base, in most instances, the oval hole serving as a door passes. The latter are posts erected in memory of the dead.

The *kexen* are generally from 30 to 50 feet in height, with a width of three feet or more at the base, and tapering slightly upwards. They are hollowed behind in the manner of a trough, to make them light enough to be set and maintained in place without much difficulty. These posts are generally covered with grotesque figures, closely grouped together, from base to summit. They include the totem of the owner, and a striking similarity is often apparent between the posts of a single village. I am unable to give the precise signification of the carving of the posts, if indeed it has any such, and the forms are illustrated better by the plates than by any description. Human figures, wearing hats of which the crowns run up in a cylindrical form, and are marked round with constrictions at intervals, almost always occur, and either one such figure, or two or three frequently surmount the end of the post. Comparatively little variation from the general type is allowed in the *kexen*, while in those posts erected in memory of the dead, and all I believe called *xat*, much greater diversity of design obtains. These posts are generally in the villages, standing on the narrow border of land between the houses and the beach, but in no determinate relation to the buildings. A common form consists of a stout, plain, upright post, round in section, and generally tapering slightly downwards, with one side of the top flattened and a broad sign-board-like square of hewn cedar planks affixed to it. This may be painted, decorated with some raised design, or to it may be affixed one of the much prized 'coppers' which has belonged to the deceased. In other cases the upright post is carved more or less elaborately. Another form consists of a round, upright post with a carved eagle at the summit. Still others, carved only at the base, run up into a long round post with incised rings at regular intervals. Two round posts are occasionally planted near together, with a large horizontal painted slab between them, or a massive beam, which appears in some instances to be excavated to hold the body. These memorial posts are generally less in height than the door posts.

The carved stone models of posts made by the Skidegate Haidas from the rock of Slate Chuck Creek are generally good representations of the *kexen*. (Several of these are figured by J.G. Swan in the publication already referred to.) Plates, flutes, and other carvings made from the same stone, though evincing in their manufacture some skill and ingenuity, have been produced merely by the demand for such things as curiosities by whites.

The use of copper, and to some extent the method of manufacturing it into various articles by hammering, has been known from time immemorial to most of the Indians of this part of the west coast. The metal has probably been for the most part obtained in trade from the Indians of the Atna or Copper River in latitude 60°17″. It is probably this familiarity with copper that has enabled the Haidas, with other tribes of the coast, so soon to acquire a proficiency in the art of working silver and iron in a rough way.

TRADITIONS AND FOLK-LORE

Of stories connected with localities, or accounting for various circumstances, there are no doubt very many among the Haidas. Of these, such as I have heard are given. The fundamental narrative of the origin of man, and the beginning of the present state of affairs is the most important of their myths. In all its minor details I believe it to be correct; that is to say, unaltered from its original traditional form. Minor shades of meaning may in some instances be indefinite, as it was obtained through the medium of the Chinook, aided by what little English my informant was master of. This, as related to me, is as follows –

Very long ago there was a great flood by which all men and animals were destroyed, with the exception of a single raven. This creature was not, however, exactly an ordinary bird, but – as with all animals in the old Indian stories – possessed the attributes of a human being to a great extent. His coat of feathers, for instance, could be put on or taken off at will, like a garment. It is even related in one version of the story that he was born of a woman who had no husband, and that she made bows and arrows for him. When old enough, with these he killed birds, and of their skins she sewed a cape or blanket. The birds were the little snow-bird with black head and neck, the large black and red, and the Mexican woodpeckers. The name of this being was *Ne-kil-stlas*.

When the flood had gone down *Ne-kil-stlas* looked about, but could find neither companions nor a mate and became very lonely. At last he took a cockle (*Cardium Nuttalli*) from the beach, and marrying it, he constantly continued to brood and think earnestly of his wish for a companion. By and bye in the shell he heard a very faint cry, like that of a newly born child, which gradually became louder, and at last a little female child was seen, which growing by degrees larger and larger, was finally married by the raven, and from this union all the Indians were produced, and the country peopled.[4]

The people, however, had many wants, and as yet had neither fire, daylight, fresh water, or the oolachen fish.[5] These things were all in the possession of a great chief or deity called *Setlin-ki-jash* who lived where the Nasse River now is. Water was first obtained in the following manner by *Ne-kil-stlas*. The chief had a daughter, and to her *Ne-kil-stlas* covertly made love, and became her accepted lover, and visited her by night many times unknown to her father. The girl began to love *Ne-kil-stlas* very much, and trust in him, which was what he desired; and at length when he thought the time ripe, he said that he was very thirsty and wanted a drink of water. This the girl brought him in one of the closely woven baskets in common use. He drank only a little, however, and setting the basket down beside him he waited till the girl was asleep, when, quickly donning his coat of feathers, and lifting the basket to his beak, he flew out by the opening made for the smoke in the top of the lodge. He was in great haste, fearing to be followed by the people of the chief. A little water fell out here and a little there, causing the numerous rivers which are now found, but on the Haida country a few drops only, like rain fell, and so it is that there are no large streams there to this day.

Ne-kil-stlas next wished to obtain fire, which was also in the possession of the same powerful being, or chief. He did not dare, however, to appear again in the chief's house, nor did the chief's daughter longer show him favour. Assuming, therefore, the form of a single needle-like leaf of the spruce tree, he floated on the water near the house, and when the girl – his former lover – came down to draw water, was lifted by her in the vessel she used. The girl drinking the water, swallowed, without noticing it, the little leaf, and shortly afterwards became pregnant, and before long bore a child who was no other than the cunning *Ne-kil-stlas*, who had thus gained an entry into the lodge. Watching his opportunity, he one day picked up a burning brand, and flying out as before by the smoke-hole at the top of the lodge, carried it away and spread fire everywhere. One of the first places where he set fire, was near the north end of Vancouver's Island, and that is the reason why so many of the trees there have black bark.[6]

All this time, however, the people were without daylight, and it was next the object of *Ne-kil-stlas* to obtain this for them. This time he tried still another plan. He pretended that he also had light, and continued to assert it, though the chief denied the truth of his statement. He, however, in some way made an object bearing a resemblance to the moon, which, while all the people were out fishing on the sea, in the perpetual night, he allowed to be partly seen from under his coat of feathers. It cast a faint glimmer across the water, which the people and *Setlin-ki-jash* thought was caused by a veritable moon. Disgusted at finding that he was not the sole possessor of light, and losing all conceit of his property, the great chief immediately placed the sun and moon where we now see them.

One thing more much desired still remained in the possession of

Setlin-ki-jash; this was the oolachen fish. Now the shag was a friend or companion of the chief, and had access to his property, including his store of oolachens. *Ne-kil-stlas* contrived that the sea-gull and the shag should quarrel, by telling each that the other had spoken evil of him. At last he got them together, when, after an angry conversation, they followed his advice and began to fight. *Ne-kil-stlas* knew that the shag had an oolachen in its stomach, and so urged the combatants to fight harder, and to lie on their backs and strike out with their feet. This they did, and finally the shag threw up the oolachen, which *Ne-kil-stlas* immediately seized. Making a canoe from a rotten log, he smeared it and himself with the scales of the oolachen, and then coming at night near the great chief's lodge, said that he was very cold, and wished to come in and warm himself, as he had been making a great fishery of oolachens, which he had left somewhere not far off. *Setlin-ki-jash* said this could not be true, as only he possessed the fish, but *Ne-kil-stlas* invited the chief to look at his clothes and at his canoe. Finding both covered with oolachen scales, he became convinced that oolachens besides those which he had must exist, and again in disgust at finding he had not the monopoly, he turned all the oolachens loose, saying, at the same time, that every year they would come in vast numbers and continue to show his liberality and be a monument to him. This they have never failed to do since that time.

This Haida story of the origin of things is substantially the same with that which I have been told by Indians of the Tinneh stock in the northern part of the interior of British Columbia. My surprise on hearing it gradually unfolded as a Haida myth was very great. It would be hazardous to theorize on the cause of this similarity of myths in tribes so distant and so dissimilar in habits, but it is certain that both its versions are derived from a common source not very remote. It may indeed be that the Haidas have adopted this story from the Tshimsians, for whose language, as we have already seen, they profess great admiration. I do not know of the existence of the story among the latter people, but they probably have it in some form, as they are supposed to be an offshoot of the great Tinneh stock of the interior country. As is always the case with these aboriginal stories, a local colouring has been given to the narrative by the Haidas, and the story of the oolachen is an addition to that which I have heard from the Tinneh. It shows the great value set upon this fish that it should receive mention among the primary necessaries of existence, such as light, water, and fire.

Ne-kil-stlas of the Haidas is represented in function and name by *Us-tas* of the Carrier Tinneh. Of *Us-tas* an almost endless series of grotesque and often disgusting adventures are related, and analogous tales are repeated about *Ne-kil-stlas*. One of these relates that he disguised himself as a dead raven, and floating on the surface of the sea was swallowed by a whale, which, by violent gripes being then induced to strand itself, became a prey to the Haidas, invisible *Ne-kil-stlas* meanwhile walking out of the whale's belly at the proper moment.

The story of the origin of the Indian tobacco referred to on a previous page, is as follows. – Long ago the Indians (first people, or ancient people – *thlin-thloo-hait*) had no tobacco, and one plant only existed, growing somewhere far inland in the interior of the Stickeen country. This plant was caused to grow by the deity, and was like a tree, very large and tall. With a bow and arrows, a man shot at its summit, where the seed was, and at last brought down one or two seeds, which he carried away, carefully preserved, and sowed in the following spring. From the plants thus procured all the tobacco afterwards cultivated sprung.

The killer whale, formerly noted as being the representative of the principle of evil, is dreaded by the Haidas, who say that these animals break canoes and drown the Indians, who then themselves become whales. The chief of the whales is the evil one himself, or his nearest analogue in the Haida mind. It is told that in the times of the grandfathers of men now living, two Haidas belonging to Klue's Village went out in a canoe to kill these whales, apparently as a daring adventure. They had paddled far out to sea when the canoe was surrounded by a great number of these evil creatures, which were about to break it in pieces. One of the men, grasping his knife, said to the other that if he drowned and became as a whale, he would still hold his knife and stab the others. The second man holding to a fragment of the canoe, floated near an island and swam ashore. The first was drowned, but his companion who had escaped, soon heard strange and very loud noises beneath the island, like great guns being fired. Presently a vast number of fish floated up dead, and with them a large whale of the malevolent kind above described. This had a great wound in its side, from which much blood flowed. The medicine-man of the village said afterwards that he knew – or saw – that the whale so killed was the chief among these creatures, and that the Indian who had killed him had now become chief in his stead.

A remarkable hill, called *Tow*, stands on the shore between Rose Point and Masset. One side is a steep cliff, while the other slopes more gradually. On the upper part of the inlet above Masset, is another hill about the same size and also precipitous on one side, called *Tow-us-tas-in* or 'Tow's Brother.' The story is that the two hills were formerly together where Tow's brother still stands, but that on one occasion Tow's brother devoured the whole of a lot of dog-fish which was in dispute between them, and that Tow being much angered went away to the open coast, where he now is.

It is also related that the summit of the hill called Tow was formerly inhabited by a very great spider, which, when a man passed, would swing itself down by its rope, catch him up, and devour him. After a time a Haida killed this spider with a spear.

Nai-koon or Rose Point (the Haida name meaning long nose) is a place full of real or imagined terrors to the Haidas. It is a dangerous and treacherous point to round at any time but in very fine weather, and many Indians have been drowned there on different occaions. They say that strange (uncanny) marine creatures inhabit its neighbourhood, and believe that if a man laugh never so little in rounding the spit, they are sure to work him evil. The father of my informant, with other Haidas in a canoe, saw one of these creatures. It was like a man, but very large, with hair hanging down to its shoulders. It raised itself out of the water to its middle, and frightened the Indians very much, but caused them no harm. Two vessels belonging to the Hudson Bay Company have been wrecked on this spit, and one of the Haida medicine-men says that the souls of these haunt the place yet. About thirty years ago a great many Indians going in canoes to profit by a dead whale that had been cast up on the spit, were drowned between Masset and that place.

There is also told in connection with Rose Point a story of a gigantic beaver. This animal, it is said, inhabits its vicinity, and when it wishes to come to the surface produces a dense fog, the water at the same time becoming very calm. The fog may, perhaps, clear away enough to allow some one watching in a retired nook to see the great beaver; but should the animal catch sight of any human being it instantly strikes the water with its tail and disappears. To laugh at the beaver, or make light of him in any way, is certain to bring bad luck; and any one seeing him must, on his return to the lodge, throw little offerings on the fire. The Tshimsians have a similar story of an immense beaver which inhabits the vicinity of Dundas Island.

FIRST CONTACT WITH EUROPEANS – FUR TRADE

During Captain Cook's last voyage in the Pacific, it was discovered that a lucrative trade in furs might be opened between the north-western coast of America and China, and though the existence of a part of the Queen Charlotte Islands had been known to the Spaniards since the voyage of Juan Perez, who was despatched by the Viceroy of Mexico in 1774, it is to the traders who followed in the track of Cook that we owe most of the earlier discoveries in the vicinity of Queen Charlotte Islands, and it is they who appear first to have come in contact with the Haidas. Before many years a number of vessels were engaged in the fur trade on this part of the west coast. Vancouver in the Notes and Miscellaneous Observations appended to his journal, states that in 1792 this trade gave employment to upwards of twenty sail of vessels, of which he gives a list, with the names of the captains. From this it would appear that five of the vessels were owned in London, one in Bristol, two in Bengal, three in Canton, six in Boston, one in New York, two in Portugal, and one in France. Most of these have left no record of their voyages, but in the published narratives of those of Dixon and Meares, already referred to, some account of the method of trade with the natives, and of their appearance, manners and customs is found.

Toward the beginning and during the earlier half of the present century, the Queen Charlotte Islands continued to be not unfrequently visited by these trading vessels, but the sea otter, the skins of which were the most valuable article of trade possessed by the islanders, having, through continuous hunting, become extremely scarce, vessels other than mere coasters have seldom called at any of the ports for many years, and our knowledge of the geography of the islands and home manners and customs of the natives has not been added to.

It is probable that La Perouse, who coasted a part of the Queen Charlotte Islands in 1786, had some intercourse with the natives, but the earliest notice of them I have been able to find is that given by 'W.B.,' the anonymous author of the letters in which the account of the voyage of the *Queen Charlotte,* of which Captain Dixon was commander, is given. He writes[7] under date of July 1st, 1787, – 'At noon we saw a deep bay,[8] which bore north-east by east, the entrance point to the northward, north-east by north; and the easternmost land south-east, about seven leagues distant. Our latitude was $54°22''$ N.; and the longitude $133°50''$ W. During the afternoon, we had light variable winds, on which we stood to the northward, for fear we should get to leeward of the bay in sight, and we were determined to make it if possible, as there was every probability of meeting with inhabitants. During the night we had light variable airs in every direction, together with a heavy swell from the south-west; so that in the morning of the 2nd we found our every effort to reach the bay ineffectual; however, a moderate breeze springing up at north-east, we stood in for the land close by the wind with our starboard tacks on board. At seven o'clock, to our very great joy, we saw several canoes full of Indians who appeared to have been out at sea, making toward us. On their coming up with the vessel, we found them to be a fishing party; but some of them wore excellent beaver[9] cloaks ... The Indians we fell in with in the morning of the 2nd of July, did not seem inclined to dispose of their cloaks, though we endeavoured to tempt them by exhibiting various articles of trade, such as toes, hatchets, adzes, howels, tin kettles, pans, &c., their attention seemed entirely taken up with viewing the vessel, which they apparently did with marks of wonder and surprise. This we looked on as a good omen, and the event showed that *for once* we were not mistaken. After their curiosity, in some measure, subsided, they began to trade, and we presently bought what cloaks and skins they had got, in exchange for toes,[10] which they seemed to like very much. They made signs for us to go in towards the shore, and gave us to understand that we should find more inhabitants, and plenty of furs. By ten o'clock we were within a mile of the shore, and saw the village where these Indians dwelt right abreast of us; it consisted of about six huts, which appeared to be built in a more regular form than any we had yet seen, and the situation very pleasant, but the shore was rocky, and afforded no place for us to anchor in. A bay now opened to the eastward, on which we hauled by the wind, which blew pretty fresh from the northward and eastward, and steered directly for it. During this time several of the people whom we traded with in the morning had been on shore, probably to show their newly acquired bargains; but on seeing us steer for the bay, they presently pushed after us, joined by several other canoes. As we advanced up the bay, there appeared to be an excellent harbour, well land-locked, about a league ahead; we had soundings from ten to twenty-five fathoms water, over a rocky bottom, but unluckily, the harbour trended right in the wind, and at one o'clock the tide set so strongly against us, that we found it impossible to make the harbour, as we lost ground every board, on which we hove the maintop-sail to the mast, in order to trade with the Indians.

'A scene now commenced, which absolutely beggars all description, and with which we were so overjoyed, that we could scarcely believe the evidence of our senses. There were ten canoes about the ship, which contained, as nearly as I could estimate, 120 people; many of these brought most beautiful beaver cloaks, others excellent skins, and, in short, none came empty-handed, and the rapidity with which they sold them, was a circumstance additionally pleasing; they fairly quarreled with each other about which should sell his cloak first; and some actually threw their furs on board, if nobody was at hand to receive them; but we took particular care to let none go from the vessel unpaid. Toes were almost the only article we bartered with on this occasion, and indeed they were taken so very eagerly, that there was not the least occasion to offer anything else. In less than half an hour we purchased near 300 beaver skins, of an excellent quality; a circumstance which greatly raised our spirits, and the more, as both the plenty of fine furs, and the avidity of the natives in parting with them, were convincing proofs, that no traffic whatever had recently been carried on near this place, and consequently we might expect a continuation of this plentiful commerce. That thou mayest form some idea of the cloaks we purchased here, I shall just observe, that they generally contain three good sea-otter skins, one of which is cut in two pieces, afterwards they are neatly sewed together, so as to form a square, and are loosely tied about the shoulders with small leather strings, fastened on each side.

'At three o'clock, our trade being entirely over, and the wind still against us, we made sail, and stood out of the bay, intending to try again for the harbour in the morning ... On the morning of the 3rd, we had a fresh easterly breeze, and squally weather, with rain; but as we approached the land it grew calm; and at ten o'clock, being not more than a mile distant from the shore, the tide set us strongly on a rocky point to the northward of the bay, on which the whaleboat and yawl were hoisted out and sent ahead, to tow the vessel clear of the rocks.

'Several canoes came alongside, but we knew them to be our friends whom we had traded with the day before, and found that they were stripped of everything worth purchasing, which made us less anxious of getting into our proposed harbour, as there was a greater probability of our meeting with fresh supplies of furs to the eastward.'

Four years later, Captain Douglas, the colleague of Meares, visited this place on his trading voyage. His people were probably the first whites to land on any part of the Queen Charlotte Islands. In the narrative of his voyage, a few details in regard to the coast and behaviour of the natives are given. From Meares' volume (p. 364) the following extracts of interest in this connexion are made. The first paragraph refers to June 19, 1789. –

'The weather was moderate and cloudy, with the wind from the south-west. At sun-set, there being the appearance of an inlet, which bore south-south-west, they stood across a deep bay, where they had irregular soundings, from twenty-six to eleven fathoms water, at the distance of two leagues from the shore; the wind dying away they dropped the stream anchor, the two points which form the bay, bearing from west, one quarter north, to north-east half east, distant from the shore four miles. It was now named McIntyre's Bay,[11] and lies in the latitude of 53°58" North, and longitude 218°6" East.

'In the morning of the 20th, the long-boat was dispatched to the head of the bay, to discover if there was any passage up the inlet; and the account received on her return was, that toward the head of the bay a bar run across, on which the long-boat got aground; but that within it there was the appearance of a large sound. Several canoes now came along-side the ship, and having purchased their stock of furs, Captain Douglas got under way to look into an inlet which he had observed the preceding year. At noon it was exceedingly hazy, and no observation was made.

'Early in the afternoon the long-boat was sent, well-manned and armed, to examine the inlet and sound for anchorage. At five o'clock they dropped the lower anchor in twenty-five fathoms water, about four miles from the shore, and two from a small barren rocky island, which happened to prove the residence of a chief, named Blakow-Coneehaw, whom Captain Douglas had seen on the coast in his last voyage. He came immediately on board, and welcomed the arrival of the ship with a song, to which two hundred of his people formed a chorus of the most pleasing melody. When the voices ceased, he paid Captain Douglas the compliment of exchanging names with him, after the manner of the chiefs of the Sandwich Islands.

'At seven in the morning (June 21st) they stood up the inlet, and at nine came to, in eighteen fathoms water, where they moored the ship[12] with the stream anchor. Through this channel,[13] which is formed by Charlotte's Islands, and an island that lies off the west end of it, the tide was found to run very rapid. The passage takes its course east and west about ten or twelve miles, and forms a communication with the open sea. It was now named Cox's Channel. Very soon after the ship was moored, the long-boat was sent to sound in the mid-channel, but no soundings could be obtained with eighty fathoms of line; but near the rocks, on the starboard shore, they had twenty and thirty fathoms water.

'Having been visited the preceding night by two canoes, which lay on their paddles, and dropped down with the tide, as was supposed, in expectation of finding us all asleep, they were desired to keep off, and finding themselves discovered they made hastily for the shore. As no orders had been given to fire at any boat, however suspicious its appearance might be, these people were suffered to retreat without being interrupted. This night, however, there happened to be several women on board, and they gave Captain Douglas to understand, that if he or his crew should fall asleep, all their heads would be cut off, as a plan had been formed by a considerable number of the natives, as soon as the lights were out, to make an attempt on the ship. The gunner therefore received his instructions, in consequence of this information, and soon after the lights were extinguished, on seeing a canoe coming out from among the rocks, he gave the alarm, and fired a gun over her, which was accompanied by the discharge of several muskets, which drove her back again with the utmost precipitation.

'In the morning the old chief Blakow-Coneehaw, made a long speech from the beach; and the long-boat going on shore for wood, there were upwards of forty men issued from behind a rock, and held up a thimble and some other trifling things which they had stolen from the ship; but when they found that the party did not intend to molest them, they gave a very ready and active assistance in cutting wood, and bringing the water casks down to the boat. Some time after the chief came on board, arrayed, as may be supposed, in a fashion of extraordinary ceremony, having four skins of the ermine hanging from each ear, and one from his nose; when, after Captain Douglas had explained to him the reason of their firing the preceding night, he first made a long speech to his own people, and then assured him that the attempt which had been made, was by some of the tribe who inhabited the opposite shore; and entreated, if they should repeat their nocturnal visit, that they might be killed as they deserved. He added, that he had left his house, in order to live along-side the ship, for the purpose of its protection, and that he himself had commanded the women to give that information which they had communicated. The old man exercised the most friendly services in his power to Captain Douglas, and possessed a degree of authority over his tribe, very superior to that of any other chief whom they had seen on the coast of America.

'In the afternoon Captain Douglas took the long-boat and ran across the channel, to an island[14] which lay between the ship and the village of Tartanee, and invited the chief to be of the party; who, having seen him pull up the wild parsley and eat it, he was so attentive as to order a large quantity of it, with some salmon, to be sent on board every morning.

'At six o'clock in the morning of the 23rd, finding the ground to be bad, they ran across the channel to a small harbour,[15] which is named Beale's Harbour, on the Tartanee side; and at ten dropped anchor in nineteen fathoms water, about half a cable's length from the shore; the land locked all round, and the great wooden images of Tartanee, bore east, one quarter north; the village on the opposite shore bearing south half west. This harbour is in the latitude of $54°18''$ North, and longitude $227°6''$ East. It was high-water there at the change, twenty minutes past midnight; and the tide flows from the westward, sixteen feet perpendicular. The night tides were higher by two feet than those of the day.

'The two following days were employed in purchasing skins, and preparing to depart; but as all the stock of iron was expended, they were under the necessity of cutting up the hatch-bars and chain plates.

'On the morning of the 27th, as soon as the chief returned, who had gone on shore the preceding evening, to get a fresh supply of provisions, Captain Douglas gave orders to unmoor, and a breeze springing up, at half-past nine they got under way, and steered through Cox's Channel, with several canoes in tow. At eleven, having got out of the strength of the tide, which runs very rapid, they hove to, and a brisk trade commenced with the natives, who bartered their skins for coats, jackets, trousers, pots, kettles, frying-pans, wash-hand basons, and whatever articles of similar nature could be procured, either from the officers or from the men; but they refused to take any more of the chain plates, as the iron of which they were made proved so brittle that it broke in their manufacturing of it. The loss of the iron and other articles of trade, which had been taken out of the ship by the Spaniards, was now severely felt, as the natives carried back no small quantity of furs, which Captain Douglas had not the means of purchasing.

'This tribe is very numerous; and the village of Tartanee stands on a very fine spot of ground, round which was some appearance of cultivation; and in one place in particular it was evident that seed had been lately sown. In all probability Captain Gray, in the sloop *Washington*, had fallen in with this tribe, and employed his considerate friendship in forming this garden; but this is a mere matter of conjecture, as the real fact could not be learned from the natives.[16] From the same benevolent spirit Captain Douglas himself planted some beans, and gave the natives a quantity for the same useful purpose; and there is little doubt but that excellent and wholesome vegetable, at this time, forms an article of luxury in the village of Tartanee. This people, indeed, were so fond of the cooking practiced on board the *Iphigenia*, that they very frequently refused to traffic with their skins, till they had been taken down to the cabin, and regaled with a previous entertainment.

Such is the first account of these Indians by the Whites. They themselves also preserve some traditions of the meeting. On asking the Chief Edensaw (*It-in-sa*) if he knew the first white man whom the Haidas had seen, he gave me, after thinking a moment, the name of Douglas, very well pronounced. Edensaw is now chief of the *Yā-tza* Village, west of Virago Sound, the *Kung* village at Virago Sound, over which he formerly presided, being nearly abandoned for the new site. Ten years or more ago, his village was on the south side of Parry Passage, but this has now been altogether given up, and the houses are rapidly crumbling away. There is little doubt that the chief with whom Captain Douglas is said to have exchanged names was a predecessor of Edensaw's, bearing, as is customary, the same name. This, with the prefix Blakow is given as Coneehaw by Douglas, and it is due to the fact of the ceremonial exchange of names having taken place, that of Douglas has been handed down to the present Edensaw, while those of Dixon and his people have been forgotten. It may generally be observed, however, that the Indians are particular in enquiring the names of whites who come among them, and it may be noted in this connection that those near the mouth of the Bella Coola River were able to give Sir Alexander McKenzie the name of Vancouver (pronounced by them Macubah) as having lately been among them, when he arrived at the coast after his celebrated journey by the Peace River.

As we have seen, however, Edensaw was wrong in saying that Douglas was the first white man seen by the Haidas, as Dixon, but two years before had been at the same spot. I did not know at the time I asked Edensaw the question, whether his reply was correct or not; and on my pressing him as to his knowledge, he admitted that he thought white men had appeared before Douglas, but he did not know their names. It was near winter, he said, a very long time ago, when a ship under sail appeared in the vicinity of North Island. The Indians were all very much afraid. The Chief shared in the general fear, but feeling that it was necessary for the sake of his dignity to act a bold part, he dressed himself in all the finery worn in dancing, went out to sea in his canoe, and on approaching the ship performed a dance (probably the Ska-ga). It would appear that the idea was at first vaguely entertained that the ship was a great bird of some kind, but on approaching it, the men on board were seen, and likened, from their dark clothing and the general sound and unintelligible character of their talk, to shags, – which sometimes indeed look almost human as they sit upon the rocks. It was observed that one man would speak whereupon all the others would immediately go aloft, till, something more being said, they would as rapidly descend. The Haidas further relate various childish stories of the surprise of those who, in a former generation, first became acquainted with many things with which they are now familiar, and profess to look upon these, their immediate predecessors, with much contempt. They say, for instance, that an axe having been given to one it pleased his fancy on account of its metallic brightness, which he likened to the skin of a silver salmon. He did not know its use, but taking the handle out, hung it round his neck as an ornament. A biscuit being given to another, he supposed it to be made of wood, and being after some time induced to eat it, finds it altogether too dry. Molasses, tasted for the first time by an adventurous Haida, pronounced very bad and his friends warned against it.

On questioning another Haida of the north part of the island, he also affirmed that the first whites had been seen near the North Island, and added that they arrived at the season when almost all the people were away at various rivers making their salmon fishery. This would be about the month of September, which agrees pretty well with Edensaw's account, and shows that the story above given cannot refer either to Douglas or Dixon, who arrived in June and July. It agrees well with the date at which Bodega and Maurelle must have passed this part of the coast on their way southward in 1775, but it appears improbable that they had any intercourse with the Haidas at this time.

VILLAGES

It is here proposed to note the various villages now inhabited by the Haidas, or of which traces still remain, beginning with those of the vicinity of North Island. It must be premised, however, that owing to the prevalent custom by which a village is spoken of by the hereditary family name of the chief, while it has besides a proper local name, and very frequently a Tshimsian equivalent for the latter by which it is also in some cases familiarly called by the Haidas themselves, much difficulty is found in correlating the villages now found with those mentioned by others.

In Parry Passage there are three village sites, two of which are on the south side, and completely abandoned. The outer or western of these shows the remains of several houses and carved posts, and is called *Kāk-oh*. The second, about half a mile further East is named *Kioo-sta*, and has been a place of great importance. This, as already mentioned, seems to have been Edensaw's place of residence at the time of Douglas' visit, and has probably been deserted for about ten years. It is nearly in the same state with the first mentioned, the houses, about twelve in number, and carved posts still standing, though completely surrounded by rank grass and young bushes, overgrown with moss, and rapidly falling into decay. It is difficult to imagine on what account this village has been abandoned, unless from sheer lack of inhabitants, as it seems admirably situated for the purposes of the natives. Many of the larger articles of property, including boxes, troughs, and other wooden vessels and stone mortars have not been removed from the houses.

On the opposite side of Parry Passage, facing a narrow channel between North Island and Lucy Island is the village which Douglas calls Tartanee. It now consists of but six houses, small and of inferior construction, and a single carved post stands a little apart from the village, but is not very old. We were informed that anciently a very large village stood here, but did not ascertain whether its inhabitants were driven away as a consequence of war with other Haidas, whether they migrated, or whether the village was simply abandoned owing to the great decrease in numbers. The present village is said to have been built after the destruction of the earlier one, a statement borne out by the fact that none of the old carved posts referred to by Douglas, and no substantial houses are now seen. There would doubtless have been propped or patched up, and thus preserved, had the spot been continuously inhabited. Douglas' account is somewhat confused, and has probably been communicated to Meares some time after the date of the events to which it relates; he mentions, however, no other chief but Blakow-Coneehaw, which would seem to show that the whole vicinity of Parry Passage was embraced in a single chieftaincy at the time of his visit.

In the first bay east of Klas-kwun Point, between North Island and the entrance of Virago Sound, the *Ya-tza*, or knife village is situated. Like many of the Haida villages, its position is much exposed, and it must be difficult to land at it with strong northerly and north-easterly winds. This village site is quite new, having been occupied only a few years. There are at present eight or ten roughly built houses, with few and poorly carved posts. The people who formerly lived at the entrance to Virago Sound are abandoning that place for this, because, as was explained to me by their chief, Edensaw, they can get more trade here, as many Indians come across from the north. The traverse from Cape Kygane or Muzon to Klas-kwun is about forty miles, and there is a rather prominent hill behind the point by which the canoe-men doubtless direct their course. At the time of our visit, in August 1878, a great part of the population of the northern portion of the Queen Charlotte Islands was collected here preparatory to the erection of carved posts and giving away of property, for which the arrival of the Kai-ga-ni Haidas was waited, these people being unable to cross owing to the prevalent fog and rough weather.

The village just within the narrow entrance to Virago Sound, from which these people are removing, is called *Kung*, it has been a substantial and well-constructed one, but is now rather decayed, though some of the houses are still inhabited. The houses arranged along the edge of a low bank, facing a fine sandy beach, are eight or ten in number, some of them quite large. The carved posts are not very numerous, though in a few instances elaborate. In J.F. Imray's North Pacific Pilot, a few notes on harbours, &c., in the Queen Charlotte Islands are given, and it is stated, in mentioning Virago Sound that the Indian village 'is to be built' inside a point on the western side of the narrowest part of the entrance. This is where the Kung village now stands. The date of the note is not given, but it is probably 1860 when the sketch map of the Sound was made.

About the entrance to Masset Inlet there are three villages, two on the east side and one on the west. The latter is called *Yān* and shows about twenty houses new and old, with thirty carved posts. The outer of these, on the east side, at which the Hudson Bay Post is situated, is named *Ut-te-was*, the inner *Kā-yung*. The *Ut-te-was* village is now the most populous, and there are in it about twenty houses, counting both large and small, with some from which the split cedar planks have been carried away, leaving only the massive frames standing. Of carved posts there are over forty in all, and these, with those of the northern part of the islands generally, show a considerable difference as compared with those of Skidegate, and other southern villages.

The styles of the northern posts are somewhat more varied, and the short, stout form, with a sign-board-like square formed of split planks at the top, is comparatively rare. Some of the Masset posts are merely stout poles, with very little carving, and at this place a thick, short post with a conical roof was observed, none like which were elsewhere seen. At the south end of the *Ut-te-was* Village is a little hill, the houses on and beyond which appear to be considered as properly forming a distinct village, though generally included in the former. The remaining Masset village (*Ka-yung*) is smaller than this one, and was not particularly examined. The principal chief of this vicinity is named *Wē-he*; he is an old man, rather stout and with nearly white hair and beard. I did not learn the precise extent of his authority, or whether, or in what degree, it may embrace the villages beyond that in which he resides.

The name Masset is of uncertain origin. Some of the natives when questioned about it, said that it has been given by the whites; while others believe that it has been extended to the whole inlet by the whites, but was the same with that of a small island which lies a little higher up the channel than the villages, and is said to be called *Maast* by the Haidas. It is unfortunate that so many places on this part of the west coast have been frequently renamed, owing to the ignorance of the names given by former explorers, but not widely published by them. The name Massette occurs, evidently denoting the place now so called in Mr. Work's table given on a following page, and constructed between 1836 and 1841. It is also found on the map illustrating Greenhow's Northwest Coast of North America, dated 1840, as Massette, but is attached to a supposed village between the positions of Masset Inlet and Virago Sound. It is suspiciously like Mazaredo, a name given by Caamano in 1793; but this, according to Greenhow's identification, is the same place known to the American traders as Craft's Sound, which is identical with Virago Sound of the modern charts, and this identification appears also to be borne out by Vancouver's chart.

A number of small houses, occupied during the summer, or salmon-fishing season, are scattered about the shores of the southern expansion of Masset Inlet. Of these, two are situated on the Ain River near its mouth, and several near the mouth of the Ya-koun. These summer houses are always small and slightly built compared with those of the permanent villages, and no attempt is made to erect any carved posts or symbols such as are appropriate at the main seat of the family.

On the north shore of Graham Island, east of Masset, and about a mile and a half from Tow Hill, is a temporary village also belonging to the Masset Indians, and occupied during the dog-fish and halibut fishery. A few small potato gardens surround the houses, which are of the unpretentious character above described, and about half a dozen in number.

Just east of Tow Hill, and on low ground on the east bank of the Hiellen River, a few much-decayed carved posts and beams of former houses are still standing, where, according to the Indians, a large village formerly existed. Its disappearance is partly accounted for by the fact that the sea has washed away much of the ground on which it stood. As the subsoil is only sand and gravel, this might easily have occurred during a single heavy storm coming from an unusual direction, or otherwise under exceptional conditions. It is probably that called *Ne-coon*, and credited with five houses in Mr. Work's table given further on. *Ne-coon* or *Nai-koon* is, however, the name of the whole northeast point of the island. North of Cape Ball, or Kul-tow-sis, on the east coast of Graham Island, the ruins of still another village yet remain. It is said to have been populous, and is near some excellent halibut banks. It is doubtless that called *A-se-guang* in Mr. Work's list, and said to have nine houses.

Tl-ell is the name of a tract of country north of the entrance to Skidegate, between Boulder Point and the mouth of a large stream twelve miles beyond it. About nine miles from Boulder Point, some posts are still standing, of an old house which must have been of great size and built of very heavy timbers. This was erected by the Skidegate chief of one or two generations back, concerning whose great size and powers many stories are current among the Haidas. The region came into the possession of Skidegate as the property of his wife, but was afterwards given by him to the Skedans of that day as a peace-offering for the wounding or killing of one of his (Skedans) women. The tract thus now belongs to Skedans, and is valued as a berry ground.

Skit-ei-get, or Skidegate Village as it is ordinarily called, situated in the inlet of the same name, and extended along the shore of a wide bay with sandy beach, is still one of the most populous Haida villages, and has always been a place of great importance. It has suffered more than most places, however, from the habit of its people in resorting to Victoria and other towns to the south. There are many unoccupied and ruinous houses, and fully one-half of those who still claim it as their residence are generally absent. The true name of the town is, I belief, *Hyohai-ka*, while *Skit-ei-get* is that of the hereditary chief. It is called *Kil-hai-oo* by the Tshimsians. There are now standing in this village about twenty-five houses, counting some of which the beams only remain, and several which are uninhabited. Of carved posts there are in all about fifty-three, making on an average two for each house, which was found also to be about the proportion in several other places. Nearly one-half of these are monumental posts or *x-at*, it being rare to find more than a single door-post or *ke-xen* for each house. Mr. Work assigns forty-eight houses to this place, which is not improbably correct for the date to which he refers, as there are signs that the village has formerly been much more extensive, and the Skidegate Haidas themselves never cease to dwell on the deplorable decrease of the population and ruin of the town. One intelligent man told me that he could remember a time – which by his age could not have been more than thirty years ago – when there was not room to launch all the canoes of the village in a single row the whole length of the beach, when the people set out on one of their periodical trading expeditions to Port Simpson. The beach is about half a mile long, and there must have been from five to eight persons in each canoe. It is not improbable that this is a somewhat exaggerated statement, but it serves to show the idea of the natives themselves as to the extent of the diminution they have suffered.

Dixon cruised northward along the east coast of the Queen Charlotte Islands about as far as Skidegate, in July, 1787, whence he turned southward for Nootka. He did not come to an anchor, but gives the following particulars, probably relating to the people of this place:[17] –

'Early in the afternoon (July 29th) we saw several canoes coming from shore, and by three o'clock we had no less than eighteen alongside, containing more than 200 people, chiefly men; this was not only the greatest concourse of traders we had seen, but what rendered the circumstance additionally pleasing was the quantity of excellent furs they brought us, our trade now being equal, if not superior to what we had met with in Cloak Bay, both in the number of skins, and the facility with which the natives traded, so that all of us were busily employed, and our articles of traffic exhibited in the greatest variety; toes, hatchets, howels, tin kettles, pewter basons, brass pans, buckles, knives, rings, &c., being preferred by turns, according to the fancy of our numerous visitants. Amongst these traders was the old chief, whom we had seen on the other side of the islands, and who now appearing to be a person of the first consequence, Captain Dixon permitted him to come on board[18] ... On our pointing to the eastward and asking the old man whether we should meet with any furs there, he gave us to understand that it was a different nation from his, and that he did not even understand the language, but was always at war with them; that he had killed great numbers and had many of them in his possession.

'The old fellow seemed to take particular pleasure in relating these circumstances, and took uncommon pains to make us comprehend his meaning; he closed his relation with advising us not to come near that part of the coast, for that the inhabitants would certainly destroy us. I endeavoured to learn how they disposed of the bodies of their enemies who were slain in battle; and though I could not understand the chief clearly enough *positively* to assert, that they are feasted on by the victors; yet there is too much reason to fear, that this horrid custom is practised on this part of the coast; [!] the heads are always preserved as standing trophies of victory.

'Of all the Indians we had seen, this chief had the most savage aspect, and his whole appearance sufficiently marked him as a proper person to lead a tribe of cannibals. His stature was above the common size; his body spare and thin, and though at first sight he appeared lank and emaciated, yet his step was bold and firm, and his limbs apparently strong and muscular; his eyes were large and goggling, and seemed ready to start out of their sockets; his forehead deeply wrinkled, not merely by age, but from a continual frown; all this, joined to a long visage, hollow cheeks, high, elevated cheek bones, and a natural ferocity of temper, formed a countenance not easily beheld without some degree of emotion. However, he proved very useful in conducting our traffic with his people, and the intelligence he gave us, and the methods he took to make himself understood, shewed him to possess a strong natural capacity.

'Besides the large quantity of furs we got from this party, (at least 350 skins) they brought several racoon cloaks, each cloak consisting of seven racoon skins, neatly sewed together; they had also a good quantity of oil in bladders of various sizes, from a pint to near a gallon, which we purchased for rings and buttons. This oil appeared to be of a most excellent kind for the lamp, was perfectly sweet, and chiefly collected from the fat of animals.'

On the following day some of the same people, in eight canoes, again came alongside, but had very few and inferior skins, their store being nearly exhausted. An attempt was made to steal some of the skins already purchased, on which several shots were fired after the offending canoe. On the day following, while endeavouring to make southward with baffling winds, the vessel was followed by a canoe containing fourteen people, who said that one of their companions had since died from a wound inflicted. No resentment was, however, shown toward the ship's company on that account, nor any fear exhibited on approaching the ship. The old chief, who seems so much to have impressed the narrator, may very probably have been the same before referred to, and described by the Haidas as of great size and striking appearance. It is unnecessary to say that no evidence of cannibalism properly so called is found among these people, though as a part of the ceremony of certain religious rites flesh was bitten from the naked arm; and in some cases it is said old people have been torn limb from limb and partly eaten, or pretended to be eaten, by several of the coast tribes. No trace now remains in the Queen Charlotte Islands of the custom of taking heads. It was formerly common on the west coast of Vancouver Island. The oil above mentioned was probably dog-fish oil, and contained in the hollow bulb-shaped heads of the gigantic sea-tangle (*Macrocystis*) of the coast.

On the west end of Maude Island, a few miles only from the Skidegate village, is now situated what may be called the New Gold Harbour Village. This has been in existence a few years only, having been built by the Haidas formerly inhabiting Gold Harbour, or Port Kuper, on ground amicably purchased from the Skidegate Haidas for that purpose. The inlet generally known as Gold Harbour, is situated on the west coast, and can be reached from Skidegate by the narrow channel separating Graham from Moresby Island. The voyage, however, includes a certain length of exposed coast, often difficult to pass in stormy weather, and the Indians, though still preserving their rights over the Gold Harbour region, and living there much of the summer, find it more convenient to have their permanent houses near Skidegate. The population of the place is about equal to that of the Skidegate village, though its appearance is much less imposing, as the houses which have been erected, are comparatively few and of small size, and there are as yet few carved posts. The two villages on the west coast, now almost abandoned by these people, are called *Kai-shun* and *Cha-atl*, – the former situated near the entrance to Gold Harbour, or *Skai-to*, the latter not far from the south-western or narrow entrance to Skidegate Channel. From one or both of these villages five canoes, with thirty-eight or forty people, came off to the *Queen Charlotte*. A few women were in the canoes, from one of whom Dixon purchased the ornamental labret which he figures in the plate opposite page 208 of his volume.

The village generally known as Cumshewa, is situated in a small bay facing toward the open sea, but about two miles within the inlet to which the same name has been applied. The outer point of the bay is formed by a little rocky islet, which is connected with the main shore by a beach at low tide. The name Cumshewa or Kumshewa is that of the hereditary chief, the village being properly called *Tlkinool*, or by Tshimsians, *Kit-ta-wās*. There are now standing here twelve or fourteen houses, several of them quite ruinous, with over twenty-five carved posts. The population is quite small, this place having suffered much from the causes to which the decrease in numbers of the natives have already been referred.

The decayed ruins of a few houses, representing a former village, which does not appear to have been large, stand just outside Cumshewa Inlet, beyond the north entrance point.

At the entrance to Cumshewa Inlet, on the opposite or south side, is the *Skedans* village, so called, as in former cases, from the chief, but of which I did not learn the proper name.[19] This is a place of more importance than the Cumshewa village proper, and appears always to have been so. Many of the houses are still inhabited, but most look old and moss-grown, and the carved posts have the same aspect. Of houses there are now about sixteen, of posts forty-four. At the time of our visit, an old woman was having a new post erected in memory of a daughter who had died some years before in Victoria. The mother having amassed considerable property for the purpose, was prepared to make a distribution when the post had been fairly put up. The village borders the shore of a semicircular bay, which forms one side of a narrow, shingly neck of land connecting two remarkable little conical hills with the main.

Klue's Village, properly called *Tanoo*, or by the Tshimsians *Lax-skīk*, is situated fourteen miles southward from the last, on the outer side of the inner of two exposed islands. The channel between the islands is so open as to afford little shelter, while the neighbourhood of the village is very rocky, and must be dangerous of approach in bad weather. There are about thirty carved posts here, of all heights and styles, with sixteen houses. The village, extending round a little rocky point, faces two ways, and cannot easily be wholly seen from any one point of view, which causes it to look less important than the last, though really possessing a larger population than it, and being in a more flourishing state than any elsewhere seen in the islands. There were a considerable number of strangers here at the time of our visit in July, 1878, engaged in the erection of a carved post and house for the chief. The nights are given to dancing, while sleep and gambling divided the portions of the day which were not employed in the business in hand. Cedar planks of great size, hewn out long ago in anticipation, had been towed to the spot, and were now being dragged up the beach by the united efforts of the throng, dressed for the most part in gaily-coloured blankets. They harnessed themselves in clusters to the ropes, as the Egyptians are represented to have done, in their pictures, shouting and ye-hooing in strange tones to encourage themselves in the work.

The *Kun-xit* Village is the most southern in the Queen Charlotte Islands. It is generally known as Ninstance or Nin-stints, from the name of the chief, and is situated on the inner side of Anthony Island of the Admiralty sketch of Houston Stewart Channel. The villages marked as occuring in Houston Stewart Channel, on the same sketch, do not exist; they have been little collections of rude houses for temporary use in summer, and have now disappeared. There are still a good many Indians here, but I have seen the place only from a distance, and know little about it. When off this place on July 23rd, Dixon was visited by eight canoes containing 'near one hundred people,' probably for the most part men, as it is mentioned, on the next day, that about 180 people, men, women and children, came out to the ship.

Besides the last mentioned, and the two villages near Gold Harbour, there were formerly two or three other places where Haidas were resident on the west coast of the islands. One of these was at Tasoo Harbour, which is reported to be a large sheet of water. I could not learn whether the village here was a permanent one, but think it must have been so. It is not improbably that designated *Too* in Mr. Work's list, and is marked on an old sketch of the islands as standing on the north-west side of the harbour. A village was situated on the island called Hippa by Dixon, of which the Haida name was, I believe, *Mustoo*. Dixon gives a sketch of the island and village in the volume already referred to. Under date July 7, 1787, he writes of this place. –

'About two o'clock in the afternoon, being close in shore, we saw several canoes putting off, on which we shortened sail, and lay too for them, as the wind blew pretty fresh. The place these people came from had a very singular appearance, and on examining it narrowly, we plainly perceived that they lived in a very large hut, built on a small island, and well fortified after the manner of an hippah, on which account we distinguished this place by the name of *Hippah Island*.

'The tribe who inhabit this hippah seem well defended by nature from any sudden assault of their enemies; for the ascent to it from the beach is steep, and difficult of access; and the other sides are well barricaded with pines and brush wood; notwithstanding which, they have been at infinite pains in raising additional fences of rails and boards; so that I should think they cannot fail to repel any tribe that should dare to attack their fortification.

'A number of circumstances had occurred, since our first trade in Cloak Bay, which convinced us, that the natives at this place were of a more savage disposition, and had less intercourse with each other, than any Indians we had met with on the coast, and we began to suspect that they were cannibals in some degree. Captain Dixon no sooner saw the fortified hut just mentioned, than this suspicion was strengthened, as it was, he said, built exactly on the plan of the hippah of the savages at New Zealand. We purchased a number of excellent cloaks, and some good skins from the Indians, for which we gave a variety of articles, some choosing toes, and others pewter basons, tin kettles, knives, &c. This tribe appeared the least we had yet seen; I could not reckon more than thirty-four or thirty-six people in the whole party; but then it should be considered that these were probably chosen men, who perhaps expected to meet with their enemies, as they were equally prepared for war or trade.'

It is possible that the 'fortified hut' seen by Dixon was a pallisaded enclosure intended for times of danger only, and not the village usually inhabited. Such a retreat formerly existed on the little island opposite Skidegate Village, though no trace of it now remains.

The last village of which I have any knowledge, stood formerly on or very near Frederick Island of the maps. Its name, or that of the island, was *Susk* or *Sīsk*. It is reputed to have been populous, but may never have been very important. Haidas belonging to this tribe came off to the *Queen Charlotte* on the 5th and 6th of July, 'bringing a number of good cloaks, which they disposed of very eagerly.' It is remarked further that: – 'These people were evidently a different tribe from that we met with in Cloak Bay, and not so numerous; I could not reckon up more than seventy-five or eighty persons alongside at one time. The furs in each canoe seemed to be a distinct property, and the people were particularly careful to prevent their neighbours from seeing what articles they bartered for.'

POPULATION OF THE QUEEN CHARLOTTE ISLANDS

As the population of the Queen Charlotte Islands has decreased, the smaller and less advantageously situated towns have been abandoned by the survivors, who have taken up their abode among the larger tribes to which they have happened to be related by marriage or otherwise. When the Indians are questioned as to why these places have been given up, they invariably say that all the people are dead, which may not be absolutely correct. Not any of the inhabited villages, however, now contain a tithe of the people for whom houses are yet standing.

It is very difficult in all cases to form estimates of the number of the aboriginal tribes when first discovered, and it is a common error, from the too literal acceptance of the half fabulous stories of the survivors, to greatly over-estimate the former population. The writer of the narrative of Captain Dixon's voyage has certainly not fallen into this mistake. He writes (p. 224): – 'The number of people we saw during the whole of our traffic, was about eight hundred and fifty; and if we suppose an equal number to be left on shore, it will amount to one thousand seven hundred inhabitants, which, I have reason to think, will be found the extreme number of people inhabiting these islands, including women and children.' It is to be remembered that Dixon not only did not anchor in any of the ports, but that most of the time he kept so far from the shore as to render it improbable that more than a small proportion of the able-bodied men of each tribe should visit the ship.

The number of sea-otter skins obtained by Dixon during the cruise about the Queen Charlotte Islands was 1821, 'many of them very fine; other furs we found in less variety here than in many other parts of the coast, the few raccoons before mentioned, a few pine-martin, and some seals, being the only kinds we saw.'

I have been so fortunate as to obtain from Dr. W.F. Tolmie the subjoined estimates of the numbers of the Haida tribes. These were made between the years 1836 and 1841 by the late Mr. John Work, and, though not framed from personal acquaintance with the Haida country, are supposed to be based on the most reliable sources, with which Mr. Work's long residence on the northern part of the coast of British Columbia had made him familiar. It is likely that even at this date the population of the islands had somewhat decreased, but in all probability not very materially. On examining the table it will be found that the villages are grouped under the common names in some instances, and that it is at times difficult to recognise what place is referred to. I have, however, endeavoured to test the table in regard to those places with which I am familiar, by comparing the relative importance of the different localities at present with that assigned to them here, and otherwise, and am persuaded that the figures are substantially correct, and probably rather an under than an over-estimate if taken to represent the population when first brought into contact with the whites.

The total number of Haidas living in the Queen Charlotte Islands, as given by Mr. Work, is 6593. The whole number of the Haida nation, including the Kai-ga-ni Haidas, 8328. The number of people assigned to each house in the Queen Charlotte Islands, according to Mr. Work's table, is found to be about thirteen, which, taking into consideration the size of the houses and manner of living, is very moderate.

The present population of the northern end of the Queen Charlotte Islands is roughly estimated by Mr. Collison, the missionary there, to number about 800. In Skidegate Inlet about 500 Haidas now remain, and are probably nearly equally divided between the two villages above described. Without referring in detail to the other villages, for which no sufficiently precise information was obtained, it is probable that the total population of the islands at the present time is from 1700 to 2000. In this estimate it is intended to include all the Haidas belonging to the islands, even those who live most of the time away from their native villages. From Skidegate Inlet and places south of it, a large proportion of the natives are always absent, generally in Victoria. From the north end of the islands comparatively few go to Victoria, while a good many resort to Fort Wrangel and other northern settlements.

The number of the people of the same stock in the southern part of Alaska, who may be classed together as Kai-ga-ni, is estimated by Mr. W.H. Dall at 300.[20]

Estimate of the Number of Haida and Kai-ga-ni Indians,
made between the years 1836 and 1841, by John Work, Esq.

Name	Men	Women	Boys	Girls	Houses
Kai-ga-ni					
You-ah-noe	68	70	44	52	18
Click-ass	98	105	102	112	26
Qui-a-hanless	30	35	42	41	3
How-a-guan	117	121	113	107	27
Shaw-a-gan	53	61	54	61	14
Chat-chee-nie	65	62	59	63	18
Totals	431	454	414	436	111
Haida					
Lu-lan-na	80	76	69	71	20
Nigh-tasis	70	69	72	69	15
Massette	630	650	589	604	160
Ne-coon	24	27	29	42	5
A-se-guang	34	31	27	28	9
Skid-de-gates	191	182	176	189	48
Cum-sha-was	80	74	63	69	20
Skee-dans	115	121	98	105	30
Quee-ah	87	79	68	74	20
Cloo	169	164	105	107	40
Kish-a-win	80	74	85	90	18
Kow-welth	131	146	145	139	35
Too	45	49	50	52	10
Totals	1736	1742	1476	1639	430

Notwithstanding the alarmingly rapid decrease of the Haida people during the century, it is not probable that the nation is fated to utter extinction. Like other tribes brought suddenly in contact with the whites, they will reach, if they have not already arrived at, a certain critical point, having passed which they will continue to maintain their own, or even to grow in numbers. As already indicated, the Haidas show a special aptitude in construction, carving, and other forms of handiwork; and it should be the endeavour of those interested in their welfare to promote their education in the simpler mechanical arts, by the practice of which they may be able to earn an honest livelihood. When the fisheries of the coast are properly developed, they will also be found of great service as fishermen; and were there a ready sale for cured fish, they might be taught so to improve their native methods as to ensure a marketable product. Saw-mills must soon spring up in the Queen Charlotte Islands to utilize their magnificent timber, and it is probable that in the course of years broad acres of fertile farms will extend where now unbroken forest stands. In such industries as these the natives may also doubtless be enlisted, but before they can be prosecuted justly the Indian title must be disposed of. This, in the case of these people, will be a matter of considerable difficulty, for as we have already seen, they hold their lands not in any loose general way, but have the whole of the islands divided and apportioned off as the property of certain families, with customs fully developed as to the inheritance and transfer of lands. The authority of the chiefs is now so small that it is more than doubtful whether the people generally would acquiesce in any bargain between the chiefs in an official capacity and the whites, while the process of extinguishing by purchase the rights of each family would be a very tedious and expensive one. The negotiations will need to be conducted with skill and care. At present, anyone requiring a spot of ground for any purpose, must make what bargain he can with the person to whom it belongs, and will probably have to pay dearly for it.

Notes

INTRODUCTION

1 Geoffrey G.E. Scudder and Nicholas Gessler, eds., *The Outer Shore* (Skidegate, BC: Queen Charlotte Islands Museum Press 1989), emphasizes the evolutionary uniqueness of the islands.
2 Swan, 'The Haidah Indians of Queen Charlotte's Islands, British Columbia,' *Smithsonian Contributions to Knowledge*, vol. 21 (Washington: Smithsonian Institution 1874).
3 Dawson to Margaret Dawson, 29 December 1870, 9 April 1876, 1 July 1880. See Dawson Family Papers, McGill University Archives, Montreal (MUA).
4 W. Shannon, 'The Late Dr. Dawson,' *The Commonwealth* (Ottawa) 1 (March 1901):50.
5 Douglas Cole and Bradley Lockner, eds. *The Journals of George M. Dawson: British Columbia, 1875-1878*, 2 vols. (Vancouver: UBC Press 1989).
6 Dawson, 'Report on the Queen Charlotte Islands,' Geological Survey of Canada, *Report of Progress for 1878-79* (Montreal: Dawson Brothers 1880).

GEORGE DAWSON'S 1878 JOURNAL

1 J. Fraser Torrance, a Montrealer who was then a British Columbia gold commissioner and who later worked for the Geological Survey of Canada.
2 Probably the veteran west coast mariner Abel Douglas (1841?-1908), who had captained the vessel used by James Richardson in his geological explorations of the Queen Charlottes a few years earlier. Douglas lived for many years in Victoria until 1906, when he moved to Seattle.
3 Margaret Ann Young (Mercer) Dawson (1830-1913) was the youngest of four daughters born to a prominent Edinburgh family. Over the strenuous objections of her parents, Margaret married J.W. Dawson on 19 March 1847 and left her native Scotland for life in British North America. Al-

though possessing a retiring nature, Margaret Dawson fulfilled admirably the arduous role of a university principal's wife and mother to five children. Deeply religious, Margaret sought to inculcate Christian values and instill a Christian faith in all her children. For an intimate picture of Margaret Dawson, see Clare Margaret Harrington, 'Grandmother from Writings of Clare Margaret Harrington,' edited and added to by Lois Winslow-Spragge, typescript, Dawson Family Papers, MUA.

4 Captain Herbert George Lewis (1828-1905) was an early citizen of Victoria who came to the Northwest Coast in 1847 with the Hudson's Bay Company. Lewis worked on various vessels, including the *Beaver* and *Otter*, before giving up his command in 1886. Later, Lewis was agent for the federal Department of Marine and Fisheries, then shipping master for the port of Victoria until his death. See John T. Walbran, *British Columbia Coast Names, 1592-1906: Their Origin and History* (Ottawa: Government Printing Bureau 1909; reprint, Vancouver: J.J. Douglas 1971, 304-5).

5 Dawson's youngest brother Rankine (1863-1913) graduated from McGill Medical School in 1882. After spending time as a medical officer for the Canadian Pacific Railway in Manitoba, he left for further training in London. For four years Rankine acted as surgeon on liners of the P & O Company before settling in London. In 1896, he married Gloranna Coats and they had one child, Margaret Rita. Always prone to depression and instability, Rankine uprooted his family and moved back to Montreal. Never achieving a permanency there, they subsequently returned to London where Gloranna left Rankine. Depressed, estranged from his family, and separated from his Montreal relatives, Rankine died in a London nursing home. For a sympathetic portrayal of Rankine, see Lois Winslow-Spragge, 'Rankine Dawson – 1863-1913,' Dawson Family Papers, MUA.

6 Bernard James Harrington (1848-1907), who married Anna Dawson in 1876, was born at St. Andrews, Lower Canada, and educated at McGill and Yale universities. Harrington was appointed lecturer in mining and chemistry at McGill University in 1871 and was on staff for thirty-six years. From 1872 to 1879 he also served with the Geological Survey of Canada. Along with a large number of scientific articles, Harrington wrote a biography of William Logan, the founder of the Geological Survey of Canada, *The Life of Sir William E. Logan* (Montreal: Dawson Brothers 1883).

7 Dawson is referring to the hopes in the Cariboo region around Barkerville that 'notwithstanding the remoteness of this district, the existence of settlement and of a considerable mining population in it, justified the attempt to develop "quartz-mining." This feeling brought about the premature quartz excitement of 1877 and 1878, which was based on exaggerated ideas of the richness of certain known lodes and on erroneous views as to the facility with which gold might be extracted from the pyritous ores which these afforded. From the collapse of this excitement, vein-mining

received a severe check.' See George M. Dawson, 'The Mineral Wealth of British Columbia with an Annotated List of Localities of Minerals of Economic Value,' in Geological Survey of Canada, *Annual Report, 1887-88*, n.s. 3 [1889]:4, pt. II, Report R, 56-7.
8 Presumably 'Map of part of British Columbia between the Fraser River and the Coast Range,' prepared to illustrate Dawson's 1876 report and published in 1878 in 'Report on Explorations in British Columbia Chiefly in the Basins of the Blackwater, Salmon, and Nechacco Rivers, and on Francois Lake,' in Geological Survey of Canada, *Report of Progress for 1876-77* (Montreal: Dawson Brothers 1878), map 120.
9 Thomas Chesmer Weston (1832-1911) had come to Canada in 1859 at the request of Sir William Logan to join the Geological Survey of Canada. Weston not only became an expert in fossil collecting but was also the survey's first librarian. He spent some thirty years with the survey. For a colourful narrative of Weston's years with the survey see his *Reminiscences Among the Rocks: In Connection with the Geological Survey of Canada* (Toronto: Warwick Bro's & Rutter 1899).
10 Casella's was a London instrument firm under the direction of father Louis P. and son Louis M. Casella. Dawson's is probably a maximum or minimum thermometer (or both) to measure temperature extremes.
11 Probably James White, draftsman with the Geological Survey of Canada.
12 Geological Survey of Canada, *Report of Progress for 1876-77*.
13 Anna Lois (Dawson) Harrington (1851-1917), Dawson's oldest sister, was also his closest friend and confidante. Even after her marriage to Bernard Harrington in 1876, Anna continued to share an intimate and rich relationship with George. They corresponded regularly and George recurringly offered assistance to his sometimes beleaguered sister who had nine children. Anna remained in Montreal for her entire adult life and eventually died of a lung tumour. See also Anna Harrington, 'Early Life at McGill by a Professor's Wife – 1867-1907,' edited by Lois Winslow-Spragge, typescript, Dawson Family Papers, MUA.
14 John Eric Harrington (1877-95) was the oldest of Bernard and Anna's children. Chronically ill, Eric never survived beyond young adulthood.
15 John Sabiston (1853?-98) was the son of John Sabiston, a prominent Nanaimo pilot from 1867 until 1896. The younger Sabiston later died of brain disease in New Westminster.
16 Senator William John Macdonald (1829-1916) had been on the same railway train across the continent as Dawson. He had come to British Columbia from Scotland in 1851 and, after working for the Hudson's Bay Company, went into business for himself. He was a member of the Vancouver Island Legislative Assembly from 1860 to 1863, of the Legislative Council of the United Colonies from 1867 to 1868, and one of the three British Columbia senators from 1871 to 1915.

17 James Richardson (1810-83) was a geologist with the Geological Survey of Canada for some thirty-six years. Richardson came to North America early in life and farmed in Beauharnois County, Lower Canada, until he joined the Geological Survey in 1846 at the urging of Sir William Logan. Richardson's expeditions to British Columbia, conducted every season from 1871 to 1875, formed the basis of his pioneering work on British Columbia coal fields. See especially his 'Report on the Coal Fields of Nanaimo, Comox, Cowichen, Burrard Inlet and Sooke, British Columbia,' in Geological Survey of Canada, *Report of Progress for 1876-77* (1878), 160-92.

18 Most likely Flora Alexandrina (1867-1924), the eldest of Senator W.J. Macdonald's three daughters. She would have been nineteen or twenty at the time. (Edythe Mary and Lilias Christiana were four and six years younger.) She married Gavin Hamilton Burns in 1890. The senator's home was Armadale, a gracious house built on twenty-eight acres at Ogden Point in the James Bay district.

19 Not in the Dawson Family Papers.

20 Albert Norton Richards (1822-97) had been appointed lieutenant governor on 20 July 1876, serving until 1881. Earlier, Richards practised law in Brockville, Upper Canada, before sitting in the House of Commons from 1872 to 1874. Richards was appointed police magistrate in Victoria in 1888.

21 Captain James Robinson and Mrs. Ada C. Robinson. Robinson was one of the early navigators of the Fraser River, employed first as master of passenger ships and later with survey ships of the Department of Public Works. He was a member of the first Legislative Assembly after the union of British Columbia and Vancouver Island.

22 Probably Miss Clara Elizabeth Dupont (1843-1923), principal of Angela College, a Victoria girls' school endowed by Baroness Burdett-Coutts.

23 William Charles (1831-1903), with Alexander Munro, was in charge of the Hudson's Bay Company's British Columbia operations. Born in Scotland, Charles joined the company in 1853 and came to Vancouver Island in 1858 where he was in charge of a number of posts. In 1874 he was appointed chief factor and moved to Victoria where he resided until retirement in 1885.

24 Mount Douglas.

25 Margaret Bruce (Eberts) Robertson was the wife of Alexander Rocke Robertson, a prominent lawyer and man of affairs. He came to Victoria in 1864 and she came after their 1868 marriage (both were from Chatham, Ontario). They were cousins, and she was the sister of D.M. Eberts, variously attorney-general, speaker, and justice in British Columbia. See 'Robertson, Alexander Rocke,' *Dictionary of Canadian Biography*, 11:756-7.

26 These were families whom Dawson knew from his 1875-6 winter in Victoria. Charles Thomas Dupont (1837-1923) was the collector of inland revenue in Victoria. Dupont, like many other British Columbians, was

actively involved in the development and promotion of mining schemes. Dr. Israel Wood Powell (1836-1915) was active in many facets of Victoria life. After graduation from McGill University in 1860, Powell came to Victoria in 1862, established a busy medical practice, and became heavily involved in politics. From 1863 he sat in the Vancouver Island House of Assembly until he lost his seat when the colony amalgamated with British Columbia. In 1872 he was appointed superintendent of Indian affairs by the Canadian government. He later profited handsomely from real estate investments in Vancouver during the 1880s. See B.A. McKelvie, 'Lieutenant Israel Wood Powell, M.D., C.M.,' *British Columbia Historical Quarterly* 9 (1947):33-54. Charles Edward Pooley (1845-1912) arrived in Victoria in 1862 and, after a few months in the Cariboo, secured a clerical position with the government, eventually becoming registrar of the British Columbia Supreme Court. He was admitted to the bar in 1877 and had a very lucrative partnership with A.E.B. Davie. Elected to the British Columbia legislature in 1882 as representative for Esquimalt, Pooley remained a member for twenty-two years. See E.O.S. Scholefield and F.W. Howay, *British Columbia from the Earliest Times to the Present*, 4 vols. (Vancouver: S.J. Clarke 1914), 4:90-1.

27 Mrs. Sarah A. Bowman (d. 1900) kept a boarding house on Yates Street in Victoria where Dawson had boarded in 1875-6. Her husband, William Gile Bowman (d. 1903), owned a livery stable.

28 Probably James Cooper Keith, a native of Aberdeen and, at this time, a ledger keeper for the Bank of British Columbia. Dawson knew him from 1875-6. Keith married a daughter of Chief Factor Roderick Finlayson and later managed the bank in Vancouver.

29 Dawson had apparently missed an entry for 26 May 1878.

30 The reference is untraceable.

31 Perhaps the most scholarly of Hudson's Bay Company employees, Alexander Caulfield Anderson (1814-84) wrote several accounts and descriptions, including *Notes on Northwestern America*, the twenty-two page reprint from the *Canadian Naturalist* which is referred to here. Dawson knew him from his 1875-6 winter in Victoria. Anderson came to the Northwest Coast during 1832 in service with the Hudson's Bay Company. Anderson retired from the company in 1854, then moved to Vancouver Island in 1858 to become collector of customs and postmaster at Victoria. Later, Anderson served on the Indian Land Commission from 1876 to 1878 and acted as dominion inspector of fisheries from 1876. Anderson was a competent writer of numerous essays and pamphlets. See 'Anderson, Alexander Caulfield,' *Dictionary of Canadian Biography*, 11:16-18.

32 Or 'iktas,' a Chinook Jargon term with various meanings for 'things' such as trade goods, merchadise, or almost any possession. Here it means 'gear.'

33 Could have been one of three species, all of which are quite similar: smooth

sun star, *Solaster endeca* (L.); Dawson's sun star, *Solaster dawsoni* Verrill; and Stimpson's sea star, *Solaster stimpsoni* Verrill.

34 Giant kelp, *Macrocystis integrifolia* Bory.
35 From the spiny dogfish, *Squalus acanthias* L.
36 See Richardson, 'Report on the Coal Fields,' 176.
37 Dawson had already concluded, in a June 1877 paper, that a great glacier had filled the entire Strait of Georgia. By 1881, largely on the basis of this 1878 trip, he believed that some of these islands were examples of moraine deposits of clay and sands from the strait glacier. See George M. Dawson, 'On the Superficial Geology of British Columbia,' *Quarterly Journal of the Geological Society of London* 34 (1878):95; and George M. Dawson 'Additional Observations on the Superficial Geology of British Columbia and Adjacent Regions,' *Quarterly Journal of the Geological Society of London* 37 (1881):279.
38 Rankine put it differently: 'As Sabiston, our bold captain, lives here, & has very recently married, I am afraid that we shall be here all morning.' See R. Dawson to J.W. Dawson, 29 May 1878, Dawson Family Papers, MUA.
39 Ballenas Islands.
40 Dawson's father, Sir John William Dawson (1820-99), was one of the most prominent figures in nineteenth-century Canadian intellectual and scientific life. After studies at Edinburgh University in the 1840s, Sir William was appointed Nova Scotia's first superintendent of education by Joseph Howe. Serving for three years, he resigned in 1853 when seeking a position at Edinburgh University. Though unsuccessful in that pursuit, Sir William was unexpectedly offered the principalship of McGill in 1855. For some forty years, until his retirement in 1893, Sir William held the principal's position at McGill; under his leadership that institution emerged as a reputable centre for teaching and research. He was made a Knight Commander of the Order of St. Michael and St. George in 1884. Sir William was always active in a wide variety of intellectual activities. He published the standard text on the geology of the Maritime Provinces in 1855, wrote countless other articles on geological and palaeontological topics, and produced several volumes exploring the relationship between religion and science. Throughout his career, Sir William was in the centre of intellectual controversy. A staunch theological conservative, he was embroiled in bitter debate due to his denunciation of Darwin's evolutionary theories, and, long after most of his geological contemporaries had abandoned the 'floating ice' theory of glaciation, Sir William unswervingly held onto the concept. For further discussion of Sir William's scientific and intellectual work, see Charles F. O'Brien, *Sir William Dawson: A Life in Science and Religion* (Philadelphia: American Philosophical Society 1971); A.B. McKillop, *A Disciplined Intelligence: Critical Inquiry and Canadian Thought in the Victorian Era* (Montreal: McGill-Queen's University Press 1979), esp. 93-134;

Carl Berger, *Science, God, and Nature in Victorian Canada* (Toronto: University of Toronto Press 1983), 38-65. Sir William's autobiography was edited by Rankine Dawson and published as the rather insipid volume, *Fifty Years of Work in Canada: Scientific and Educational* (London: Ballantyne, Hanson 1901). Additional biographical data is found in 'Dawson, Sir John William,' *Dictionary of Canadian Biography*, 12:230-7; and in Henry M. Ami, 'Sir John William Dawson: A Brief Biographical Sketch,' *American Geologist* 26 (1900):1-48, which also includes a bibliography of Sir John William Dawson's writings.

41 The Hudson Bay Company's *Otter* was built in 1852 in Blackwell, England, and arrived at Fort Victoria in July 1853, replacing the *Beaver* in the coastal run. She registered 219 tons and was powered by two steam engines. The vessel was sold for scrap in San Francisco in 1890.

42 The Euclataw, Lekwitoq, or Lekwiltok were a Kwakiutl group who controlled the narrow waters of Discovery Passage at Cape Mudge and raided travellers using the inside waterway.

43 Western red cedar, *Thuja plicata* Donn.

44 East Thurlow Island.

45 Newcastle Ridge.

46 Helmcken Island.

47 Alden Westley Huson (1832?-1913) was born in New York and came from California to the Cariboo in 1858. In the early 1860s he settled on the British Columbia coast, where he operated a trading schooner along the east side of Vancouver Island, opened a store at Suquash, and then in 1881 went into the cannery business with S.A. Spencer and Thomas Earle at Alert Bay. Huson later sold his share of the cannery and worked a stone quarry. See his obituary in the *Victoria Daily Colonist*, 7 January 1913.

48 A common name for an Indian village.

49 Eulachon, *Thaleichthys pacificus* (Richardson). For a brief account of aboriginal Northwest Coast eulachon fishing, see H.A. Collison, 'The Oolachen Fishery,' *British Columbia Historical Quarterly* 5 (1941):25-31.

50 A division of the Kwakiutl.

51 Unidentifiable; his 'place of abode' would have been near today's Suquash, between Port Hardy and Port McNeill.

52 Gavin Hamilton (1835-1909) was the Hudson's Bay Company factor at Fort St. James on Stuart Lake. Hamilton arrived in Victoria in 1853 and entered the service of the Hudson's Bay Company at Fort Langley. After briefly leaving the company in 1857, Hamilton returned until 1879, when he retired and erected a sawmill and grist mill on the Cariboo Road. See also Gavin Hamilton, 'Reminiscences,' MG 29, vol. 7, National Archives of Canada (NAC). Dawson had met him at Fort St. James in July of his 1876 field season.

53 Dawson's photographs are held in the National Archives of Canada. An-

notations give their date and inventory numbers. They are listed in full in Douglas Cole and Bradley Lockner, eds., *The Journals of George M. Dawson*, 2:558-63. These photographs are: 2 June 1878, GSC221-C1 (PA 152382), GSC222-C1 (PA 152384).

54 Nimpkish Lake.
55 'Menzies Spruce' was a Sitka spruce, *Picea sitchensis* (Bong.) Carr.; 'Hemlock' was a western hemlock, *Tsuga heterophylla* (Ruf.) Sang.; and 'Cedar' was a western red cedar, *T. plicata*.
56 An echinoderm of the Class Stelleroidea, which is impossible to identify precisely.
57 Actually 'Cheap,' a version of 'chief.' Cheap, or *Kup'quhila*, was a chief at Nawitti, and, when described by Franz Boas from his 1886 visit, was then over sixty. See Boas, 'The Houses of the Kwakiutl Indians, British Columbia,' *Proceedings of the United States National Museum* 11 (1888):206-7; Boas, 'The Indians of British Columbia,' *Popular Science Monthly* 32 (1888):632-3. He enters Dawson's narrative again on 14 and 15 September.
58 The 'Porpoises' could have been either the Dall porpoise, *Phocoenoides dalli* (True), or the harbour porpoise, *Phocoena vomerina* Gill; the 'sea birds' are impossible to identify; and the 'sea-lion' was a northern sea-lion, *Eumetopias jubata* (Schreber).
59 Northern fur-seal, *Callorhinus ursinus cynocephalus* (Walbaum).
60 Halibut, *Hippoglossoides* spp.
61 Probably Walbran Island.
62 A Northern Kwakiutl, Haisla-speaking group, living on Douglas Channel.
63 The whales are impossible to identify precisely.
64 The 'bright green' anemone was probably the giant green anemone, *Anthopleura xanthogrammica* (Brandt); and the other with 'plumose tentacles milk white,' was possibly the frilled anemone, *Metridium senile* (L.).
65 Joseph William McKay (1829-1900) who was a long-time Hudson's Bay Company employee in British Columbia. In the early 1850s, McKay led explorations which established the coal reserves at Nanaimo. Later appointed a chief factor, McKay spent time in Kamloops and Victoria before his dismissal in 1878. McKay took the position of Indian agent at Kamloops, then assistant Indian superintendent for British Columbia. See 'McKay, Joseph William,' *Dictionary of Canadian Biography*, 12:641-3.
66 See 30 August 1878, for McKay's 'Chimseyan Lode.' Lieutenant Colonel Charles Frederick Houghton (1839-98) was a prominent pioneer settler and politician. Houghton came to British Columbia in 1863 and obtained all his military land grant in the north Okanagan Lake region. After building up considerable land holdings, Houghton was elected to the House of Commons in 1871. Appointed deputy adjutant-general of militia for British Columbia in 1873, Houghton later served in the Riel Rebellion and

was deputy adjutant-general in Montreal. He retired in 1896. See Margaret A. Ormsby, 'Some Irish Figures in Colonial Days,' *British Columbia Historical Quarterly* 14 (1950):76 and 'Houghton, Charles Frederick,' *Dictionary of Canadian Biography*, 12:449-51.

67 The Hudson's Bay Company had taken over the Bella Bella trading post on the site of its own Fort McLoughlin (1833-43) from an independent trader in 1870.

68 Boston was one of several Heiltsuk or Bella Bella chiefs of the 1830s. Walbran describes him as 'sharper and shrewder' than the others. His village was at the head of Lizzie Cove, Lama Pass. See Walbran, *British Columbia Coast Names*, 46. O.C. Hastings photographed the memorial a year later; his photograph is in the Royal British Columbia Museum, no. PN2404.

69 A flat sheet of copper cut and bent into a distinctive shape. See Dawson's comments in 'On the Haida Indians of the Queen Charlotte Islands,' 128. He is probably in error in asserting that they are of native metal since all extant specimens are European copper.

70 The Heiltsuk or Bella Bella had a reputation for canoe-building. The canoe for the 1876 United States Centennial Exposition in Philadelphia was purchased at Alert Bay in June 1875 by James G. Swan, acting for the United States Centennial Commission, after which it became the property of the United States National Museum. The canoe which Dawson saw may well be the one later collected by Indian Superintendent Israel W. Powell on commission for Heber R. Bishop and New York's American Museum of Natural History and now a feature of that museum's South Entrance.

71 Probably butter clams, *Saxidomus giganteus* (Deshayes).

72 Chinook Jargon for 'friends.'

73 Ratfish, *Hydrolagu colliei* (Lay and Bennett). An 'elasmobranch' is a fish belonging to the Elasmobranchii, the group of cartilaginous fishes comprising the sharks and rays.

74 McKay's Hebrew Mine was at Neekas Inlet on the north side of Spiller Channel. 'Some work was done several years ago on a vein containing pyrrhotite or magnetic iron-pyrites and copper pyrites. This deposit is known as the "Hebrew Mine," and appears to deserve further examination as a copper-ore.' See Dawson, 'The Mineral Wealth of B.C.,' 153.

75 Possibly John Williams, listed in the 1881 census as a 'mariner.' Born in Wales and residing in Victoria, he was, at the time of enumeration, thirty-six years old.

76 Unidentifiable. Rankine added that 'our cook is going to be quite a success I think. He cooks very well & keeps things beautifully clean, besides wasting very little.' See Rankine Dawson to Margaret Dawson, 8 June 1878, Dawson Family Papers, MUA.

77 William Bell Dawson (1854-1944), Dawson's brother, also became a promi-

nent scientist, even though overshadowed for many years by George and his father. William graduated from McGill University in 1874 with his Bachelor of Arts, obtained a bachelor's degree in applied science the year after, then went to Paris to attend the prestigious Ecole des Ponts et Chaussées. Following a three-year course, William returned to Canada and, after he unsuccessfully applied for several positions, went into a private engineering practice. In 1882, he joined the Dominion Bridge Company as an engineer and stayed until 1884, when he took a position as assistant engineer for the Canadian Pacific Railway Company. William spent ten years with the firm designing bridges for the many new lines being constructed. He took up, in 1884, what he considered his main professional undertaking: director of the Dominion Survey of Tides and Currents. For thirty years, until his retirement in 1924, he recorded and mapped tides and currents in the harbours and on the major steamship routes of the Canadian coasts. Upon retiring, he spent much energy writing articles and tracts proving the harmony of science with religion, a task not unlike that earlier done by his father. William married Florence Jane Mary Elliott (1864?-1945), and the couple had three sons and a daughter. See William Bell Dawson's obituary in the *Montreal Gazette*, 22 May 1944.
78 Presumably William Dawson was seeking employment as an engineer.
79 Day Point.
80 Rose Harbour.
81 12 June 1878, GSC223-C1 (PA 51097) and GSC224-C1 (PA 152383).
82 'The villages marked as occuring in Houston Stewart Channel,' Dawson noted, 'do not exist; they have been little collections of rude houses for temporary use in summer, and have now disappeared.' See Dawson, 'On the Haida,' 161.
83 Kerouard Islands.
84 Presumably Moore Head is meant.
85 Renamed Garcin Rocks in 1948.
86 Most likely Hornby Point and Quadra Rocks.
87 The 'Seals' were probably hair or harbour seals, *Phoca vitulina richardi* (Gray); the 'eagles,' bald eagles, *Haliaeetus leucocephalus* (L.); 'black & white Guillamettes,' pigeon guillemots, *Cepphus columba* Pallas; and the 'black bird with long bright red bill,' the black oystercatcher, *Haematopus bachmani* Audubon.
88 The 'Acorn shells' were possibly shells of the giant acorn barnacle, *Balanus nubilis* Darwin; the 'large mussels,' California mussels, *Mytilus californianus* Conrad; and 'Lepas,' goose barnacles, *Lepas* spp. The 'large urchins' were possibly the giant red sea urchin, *Strongylocentrotus franciscanus* (Agassiz); the 'Sea anemones' are impossible to identify; and 'starfish,' one of several species such as the common starfish, *Pisaster ochraceus* (Brandt).
89 The present Rose Inlet; Rose Harbour is now the name of Dawson's 'Snug

bight' of 12 June on the south shore of Houston Stewart Channel.
90 Cape Fanny, at the northwest entrance to the channel.
91 Anthony Island is the location of Ninstints or SgA'ngwā'-i village. See John R. Swanton, *Contributions to the Ethnology of the Haida*, Jesup North Pacific Expedition, vol. 5, pt. 1, American Museum of Natural History Memoir, vol. 8, pt. 1 (New York 1905), 277. Dawson calls it Kun-xit village, which is actually a lineage name. The village was abandoned after 1884. See Dawson, 'On the Haida,' 161. As the best preserved Haida village, it is now a World Heritage Site.
92 Like rancherie, a common term for an Indian village.
93 The 'byezoons' were moss animals of the phylum Bryzoa or Ectoprocta; the 'Corals,' one of several species such as the orange cup coral, *Balanophyllia elegans* Verrill; and the 'brachiopod,' the common Pacific brachiopod, *Terebratalia transversa* (Sowerby).
94 Coiled cephalopods, highly diversive in structure and prolific in species and genera.
95 Haida from Chaatl and Kaisun, villages associated with 'Gold Harbour,' the waters adjacent to Kuper Island on the west coast, moved to Haina or New Gold Harbour on Maude Island near Skidegate between 1870 and 1875. The ground for the new village, Dawson noted, was 'amicably purchased from the Skidegate Haidas.' See Dawson, 'On the Haida,' 160.
96 A sculpin of the family Cottidae.
97 Named by Dawson, Carpenter Bay, after the English naturalist, William Benjamin Carpenter.
98 Named South Cove by Dawson.
99 Francis Poole was a mining engineer who investigated copper reports in 1862 on behalf of the Queen Charlotte Mining Company of Victoria. He abandoned his mines in 1864 after shafts had been sunk on Burnaby and Skincuttle islands. Dawson thought his book, Poole, *Queen Charlotte Islands: A Narrative of Adventure in the North Pacific*, ed. John W. Lyndon (London: Hurst and Blackett 1872), 'chiefly remarkable for the exaggerated character of the accounts it contains.' See George M. Dawson, 'Report on the Queen Charlotte Islands 1878,' in Geological Survey of Canada, *Report of Progress for 1878-79* (Montreal: Dawson Brothers 1880), Report B, 17.
100 Poole, *Queen Charlotte Islands*, 162, notes a deposit of 'magnetic iron ore' east of Harriet Harbour.
101 Huston Inlet.
102 Copper Islands.
103 Poole Point, Burnaby Island.
104 On the southeast tip of Burnaby Island between Bluejay and Pelican coves.
105 Probably Swan Bay.
106 Dawson mistakenly wrote 'July' rather than 'June.'

107 Bag Harbour and Island Bay, both named by Dawson.
108 The 'spruce' was possibly the Douglas-fir, *Pseudotsuga menziesii* (Mirbel) Franco, but Dawson was more likely referring to the Sitka spruce, *P. sitchensis*; 'hemlock,' western hemlock, *T. heterophylla*; 'cedar,' western red cedar, *T. plicata*; and 'alder' was probably the red alder, *Alnus rubra* Bong.
109 Probably the common polypody or licorice fern, *Polypodium vulgare* L.
110 26 June 1873, GSC226-C (PA 51091).
111 Micrometer telescope or theodolite.
112 Island Bay.
113 28 June 1878, GSC227-C2 (PA 152372) and GSC228-C2 (PA 152370).
114 This became Section Cove. The exposed rocks are discussed in Dawson, 'Report on the Queen Charlotte Islands,' 55, fig. 2 and 56.
115 29 June 1878, GSC229-C2 (PA 152371) and GSC230-C2 (PA 152362).
116 Scudder Point, named by Dawson.
117 Meaning the islands, his Copper Islands, off Skincuttle Inlet, of which Skincuttle Island is one.
118 A seaweed, *Laminaria* spp.
119 The reference is to Scudder Point. In his 'Report on the Queen Charlotte Islands,' 21, Dawson terms the house 'strongly built.' The site is Skwaikun on the C.F. Newcombe map in Swanton, *Ethnology of the Haida*, opp. 277.
120 Juan Perez Sound, so named by Dawson.
121 The village of T!anū' or Tanu, but often called Clue's or Klue's village after the hereditary name of the Chief Xe-ū'. See 7 July 1878.
122 A sculpin of the family Cottidae.
123 Skaat Harbour.
124 This is difficult to follow; the best construction seems to be that Dawson did not realize that this should really be considered a part of Juan Perez Sound and not another opening.
125 1 July 1878, GSC231-C2 (PA 51089) and GSC232-C2 (PA 152363).
126 De la Beche Inlet, named by Dawson after the geologist Sir Henry De la Beche.
127 Dawson mentions only Werner Bay and Hutton Inlet in his 'Report on the Queen Charlotte Islands,' 21. These, he wrote, 'owing to the short time at my disposal, and comparatively uninteresting character of the rock sections, were not examined to their heads.'
128 Most likely Murchison Island.
129 Hotspring Island, named by Dawson. The site on the southwest side of the island was 'easily recognized by a patch of mossy green sward which can be seen from a considerable distance. Steam also generally hovers over it. The actual source of the water is not seen, but is probably not far from the inner edge of the mossy patch.' See Dawson, 'Report on the Queen Charlotte Islands,' 22. He had no thermometer to measure the heat of the warmest streams.

130 Salal, *Gaultheria shallon* Pursch.
131 Hardened rocks composed of clay minerals.
132 4 July 1878, GSC235-C2 (PA 51103) and GSC236-C2 (PA 51099).
133 Dawson is now in Darwin Sound.
134 Longfellow Creek.
135 4 July 1878, GSC237-C2 (PA 51100).
136 Through Richardson Pass into Richardson Inlet.
137 Darwin Sound.
138 Yellow cedar, *Chamaecyparis nootkatensis* (Lamb.) Spach.
139 Bent Tree Point.
140 Klunkwoi Bay, so named by Dawson.
141 Part of the San Cristoval Range. The two highest peaks west of the inlet are Apex Mountain and Mount Laysen.
142 This name has been retained.
143 De la Beche Inlet.
144 Redtop Mountain.
145 6 July 1878, GSC240-C2 (PA 152367)
146 Richardson Island.
147 Laskeek Bay.
148 Clue village, sometimes called Laskeek, but more properly Tanu. It is described with Dawson's photographs, in George F. MacDonald, *Haida Monumental Art: Villages of the Queen Charlotte Islands* (Vancouver: UBC Press 1983), 88-100. Dawson himself wrote: 'The channel between the islands is so open as to afford little shelter, while the neighbourhood of the village is very rocky, and must be dangerous of approach in bad weather. There are about thirty carved posts here, of all heights and styles, with sixteen houses. The village, extending round a little rocky point, faces two ways, and cannot easily be wholly seen from any one point of view, which causes it to look less important than the last [Skedans], though really possessing a larger population than it, and being in a more flourishing state than any elsewhere seen in the islands.' See Dawson, 'On the Haida,' 161.
149 Variously Kloo, Klew, Klue, Clew, and Cloo from one of his highest names Xeū' (meaning 'the southeast'). He often went by Kitkane or Kitkune (Gitku'n, according to Swanton, a Tsimshian word). See Swanton, *Ethnology of the Haida*, 96, 278. Both names were hereditary. He was a young man, having succeeded his uncle of the same name on the latter's death in 1876. See James G. Swan, typescript diary, 56-7, University of Washington Archives, Seattle.
150 For photographs and a description of Haida houses, see Margaret R. Blackman, *Window on the Past: The Photographic Ethnohistory of the Northern and Kaigani Haida*, National Museum of Man, Mercury Series, Canadian Ethnology Service Paper, no. 74 (Ottawa: National Museums of Canada 1981), esp. 113-35.

151 Siwash, deriving in Chinook Jargon from the French *sauvage*, had now become a derogatory term.
152 Or *'tyee,'* Chinook Jargon for 'chief.'
153 Haida from Skidegate village on Skidegate Inlet.
154 Dawson commented further on the popularity of stick-game gambling in 'On the Haida,' 122-3. For a fuller treatment of the game, see Swanton, *Ethnology of the Haida*, 58-9.
155 Lyell Island, named by Dawson after Sir Charles Lyell.
156 Tar Rock. See Kathleen E. Dalzell, *The Queen Charlotte Islands*, vol. 2, *Of Places and Names* (Prince Rupert, BC: Dalzell Books 1973), 208.
157 9 July 1878, GSC242-C2 (PA 37753).
158 Kunga Island, named by Dawson.
159 Beds of limestone consisting of layers from one to ten centimetres in thickness.
160 A major village to the north on Cumshewa Inlet.
161 Perhaps Selwyn Inlet and Talunkwan Island are meant.
162 Dana Inlet, named by Dawson after geologist James Dana.
163 Probably Logan Inlet, named after Sir William Logan.
164 Rockfish Harbour, so named by Dawson.
165 It is Tanu House 5 of MacDonald, *Haida Monumental Art*, 93, which includes a history and description of the house, along with Dawson's photographs.
166 Or rockfish, *Sebastodes* spp.
167 Skedans or Koona village (Swanton's Q!ō'na). Dawson described it as being more important than Cumshewa. 'Many of the houses are still inhabited, but most look old and moss-grown, and the carved posts have the same aspect.' See Dawson, 'On the Haida,' 160. See also 18 July 1878. It is fully described in John and Carolyn Smyly, *Those Born at Koona: The Totem Poles of the Haida Village Skedans Queen Charlotte Islands* (Saanichton, BC: Hancock House 1973); and MacDonald, *Haida Monumental Art*, 78-86.
168 Chief Skedans or Skedance, chief of Koona. The name was hereditary.
169 Indians usually collected testimonials from Europeans and showed them to visitors, even though these sometimes contained unflattering remarks.
170 There is no record of an *Olympia* being wrecked on the Queen Charlotte Islands at such a time, but the incident sounds similar to that of the *Susan Sturgis* pillage, though it was Edenshaw of the Massets who was then the dubious hero. See 23 August 1878.
171 Like other Northwest Coast languages, Haida contains no sound equivalent to the English 'r.'
172 McCoy Cove. Dawson uses the name McCoy in his text, but corrects it to McKay in his 'Report on the Queen Charlotte Islands,' 28. Captain Hugh McKay established a base and trading post at this point in 1869, but it was abandoned about six years later. According to Kathleen E. Dalzell, McKay's

Cove was renamed McCoy Cove in 1921 after a fireman on board the *Lillooet*. See Dalzell, *The Queen Charlotte Islands*, vol. 2, 268.
173 Chinook Jargon, '*illahee*' meaning country, land, or earth.
174 The 'Black bears' were American black bears, *Ursus americanus carlottae* Osgood; 'Marten,' marten, *Martes americana nesophila* (Osgood); 'Otter,' the river otter, *Lutra canadensis periclyzomae* Elliot; and the 'mouse,' probably a deer mouse, *Peromyscus maniculatus keeni* (Rhoads).
175 Dawson was correct in that no snakes were indigenous to the Queen Charlotte Islands, while the northwestern toad, *Bufo boreas boreas* Baird & Girard, was found there.
176 16 July 1878, GSC243-C2 (PA 37752); GSC244-C2 (PA 38151); and GSC245-C2 (PA 38150).
177 Doubtless his Boat Cove.
178 17 July 1878, GSC247-C2 (PA 51105).
179 Probably from stone sheep, *Ovis dalli stonei* Allen, which are found only on the mainland north of the Skeena and Nass rivers. The horn was steamed into a spoon shape and the narrow handle usually carved.
180 The Haida had begun growing potatoes early in the nineteenth century. See J.H. Van Den Brink, *The Haida Indians: Cultural Change Mainly between 1876-1970* (Leiden: E.J. Brill 1974), 34-5. In his 'On the Haida,' 107, Dawson wrote: 'The potato, called *skow-shīt* in Haida, introduced by some early voyagers, now forms an important part of the food supply. A Skidegate Indian told me that it was first grown at Skidegate, but I do not know how far this statement may be reliable.'
181 Only four photographs remain: 18 July 1878, GSC248-C2 (PA 37754); GSC249-C2 (PA 44329); GSC250-C2 (PA 38148); and GSC251-C2 (PA 38153).
182 Skedans Bay.
183 Some of Dawson's purchases are illustrated in his 'On the Haida.' Most of the artifacts he collected were placed on loan to McGill College and are now in the McCord Museum.
184 These are grave or mortuary posts, the boards holding the remains of the dead.
185 The pole is the 'Bear and Copper Mortuary.' See no. 17c of John and Carolyn Smyly, *Those Born at Koona*, 89; and Memorial 10x1 of MacDonald, *Haida Monumental Art*, 82. The box front is now in the Prince Rupert Museum, without its copper.
186 Hamilton Moffat (1832-94) joined the Hudson's Bay Company in London in 1849 and the following year was posted to Victoria. He later served at Fort Rupert and Fort Simpson before retiring in 1872. The following year he was appointed to the Department of Indian Affairs in Victoria. See Walbran, *British Columbia Coast Names*, 340-1.
187 Swanton identifies the village as St!awā's and the chief as Gō'msiwa. See his *Ethnology of the Haida*, 273. Dawson commented: 'The village generally

known as Cumshewa, is situated in a small bay facing toward the open sea, but about two miles within the inlet to which the same name has been applied. The outer point of the bay is formed by a little rocky islet, which is connected with the main shore by a beach at low tide. The name Cumshewa or Kumshewa is that of the hereditary chief, the village being properly called *Tlkinool*, or by Tshimsians *Kit-ta-wās*. There are now standing here twelve or fourteen houses, several of them quite ruinous, with over twenty-five carved posts. The population is quite small, this place having suffered much from the causes to which the decrease in numbers of the natives have already been referred.' See Dawson, 'On the Haida,' 160. For a further description, see MacDonald, *Haida Monumental Art*, 67-74.
188 These would be Southern Tsimshian from Kitkatla village near the mainland. Eulachon were abundant in Tsimshian territory but absent from the Queen Charlotte Islands.
189 For a similar description of the trade, see Dawson, 'On the Haida,' 129-30.
190 Eulachon ran on many mainland rivers, but the most abundant catch was on the Nass River. The numerous trading trails across the mountains and into the interior were customarily called 'grease trails' because of their frequent use as trading routes for the highly valued eulachon oil.
191 As Dawson later realized, Cumshewa Inlet does connect with Selwyn Inlet to the south, creating Louise Island.
192 From Sewell Inlet to Tasu Sound on the west coast of Moresby Island.
193 Kuper Inlet.
194 Dawson here realized that Cumshewa Island, a small barren rock, was not what he meant. The island referred to is Kingui Island.
195 Cumshewa Rocks.
196 Skedans Islands.
197 Fairbairn Shoals.
198 McLellan Island.
199 Mathers Creek.
200 Probably Copper Bay.
201 Skidegate Mission on the Graham Island shore of Skidegate Inlet.
202 Part of Skidegate Plateau.
203 Now Skidegate Landing in Sterling or Skidegate Harbour.
204 J. McB. Smith, together with William Sterling, had established an oilworks to extract oil from dogfish livers in 1876. His foreman and perhaps partner was Andy McGregor, presumably Dawson's 'Mcfarlane.' See Kathleen E. Dalzell, *The Queen Charlotte Islands, 1774-1966* (Terrace, BC: C.M. Adam 1963), 107-8; Johan Adrian Jacobsen, *Alaska Voyage, 1881-1883: An Expedition to the Northwest Coast of America*, trans. Erna Gunther from the German of Adrian Woldt (Chicago: University of Chicago Press 1977), 20; and Newton H. Chittenden, *Official Report of the Queen Charlotte Islands for the*

Government of British Columbia (Victoria: Printed by the Authority of the Government 1884), reprinted as *Exploration of the Queen Charlotte Islands* (Vancouver: Gordon Soules 1984), 73-4.

205 Unidentifiable.
206 This account was published, almost verbatim, by Dawson in his 'The Haidas,' *Harper's New Monthly Magazine* 45 (1882):405-6. The draft is in file 69 of the G.M. Dawson Papers, MUA. In his published report, Dawson writes of six kinds of Haida dancing ceremonies and calls this one, in the Skidegate dialect, *Kwai-o-guns-o-lung*. See Dawson, 'On the Haida,' 127-80. The term may be the same as Swanton's *Q!ā'igî'lgañ*, which he labels as a song and as part of the Wāgał potlatch, a Haida equivalent of the Cannibal Dance. See Swanton, *Ethnology of the Haida*, 162-70, esp. 166.
207 Could have been from either the horned puffin, *Fratercula corniculata* (Naumann), or the more common tufted puffin, *Lunda cirrhata* (Pallas).
208 Presumably the famous Chilkat blankets, made almost exclusively by the Chilkat Tlingit of Alaska.
209 Northern abalone, *Haliotis kamtschatkana* Jonas.
210 26 July 1878, GSC253-C2 (PA 38152); GSC254-C2 (PA 37755); and GSC255-C2 (PA 37756).
211 Jewell and Torrens islands.
212 Dawson noted that it was 'reported by the Indians that a well-marked coal seam occurs about fourteen miles from the original locality in a southeasterly direction on the south shore of Skidegate Channel [Inlet].' See Dawson, 'Note on the General Mines and Minerals of Economic Value of British Columbia with a List of Localities,' in Geological Survey of Canada, *Report of Progress for 1876-77*, 120.
213 Possibly the chub mackerel, *Scomber japonicus* Houttuyn.
214 Sandilands Island.
215 Immediately northwest of Anthracite Point, Long Inlet.
216 As Dawson was correctly informed, this would be a doctor's or shaman's interment. The body of such personages was often placed in a grave house with their ritual accessories at sites away from the villages. Dawson makes clear the distinctive burial of shamans and describes it in detail in 'On the Haida,' 116-18. For a recent examination of Haida mortuary practices, see George F. MacDonald, *Haida Burial Practices: Three Archaeological Examples*, National Museum of Man, Mercury Series, Archaeological Survey of Canada Paper, no. 9 (Ottawa: National Museums of Canada 1973), 1-5.
217 Long Inlet.
218 Located opposite the Cowgitz Coal Mine. See 30 July 1878.
219 Devil's-club, *Oplopanax horridum* (J.E. Smith) Miq.
220 Coal was first reported on Skidegate Channel by William Downie in 1859. The Queen Charlotte Coal Company was organized in 1865 by a group of Victoria merchants who obtained a Crown lease. With evidence of good

quality anthracite, the Cowgitz Coal Mine began with ample funding and optimistic prospects. Tunnels were driven and tramways constructed to tidewater, a wharf and housing for workers built, and screens and other equipment erected. Various problems, however, prevented any coal shipments until Christmas 1870 when the first 435 tons were shipped. Though mining continued, the seams proved very difficult and extraction of the coal tedious. The final blow came with the loss of a ship loaded with coal bound for San Francisco. See Dalzell, *The Queen Charlotte Islands, 1774-1966*, 69, 105-7. For technical descriptions of the seams and mine, see James Richardson, 'Report on Coal-Fields of Vancouver and Queen Charlotte Islands, with a Map of the Distribution of the Former,' in Geological Survey of Canada, *Report of Progress 1872-3* (Montreal: Dawson Brothers 1873), 57-62; and Dawson, 'Report on the Queen Charlotte Islands,' 71-7.

221 The 'Salmon berry bushes' were salmonberry, *Rubus spectabilis* Pursh; and 'fire-weed,' fireweed, *Epilobium angustifolium* L.
222 West Narrows.
223 Buck Channel.
224 A continuation of Skidegate Channel. Cartwright Sound is to the north.
225 Called North Arm by Dawson, later renamed Trounce Inlet.
226 Or '*Tenas plaklie*,' Chinook Jargon for 'evening.'
227 The 'aqueous' rocks are those produced by the action of water; and 'agglomerate' rocks are pyroclastic rocks consisting of coarse fragments and finer material.
228 William H. Woodcock, who established Woodcock's Landing on the Skeena. According to Walbran, he died in Victoria in 1877, but Newton Chittenden met him in 1882 at Fort Wrangel, where he had been for several years.
229 Douglas Inlet.
230 The reference here is to the Gold Harbour or Mitchell Inlet gold discoveries. Reportedly discovered by the Haida in about 1850, the deposits were explored by the Hudson's Bay Company in the spring of 1851 when John Work led an expedition to the area. Work found nothing, but a later party headed by Captain William Mitchell on the *Una* discovered quartz veins with gold. The ore recovered by Mitchell was lost when the *Una* was wrecked off Neah Bay on the Washington coast. Another season's work almost exhausted the small vein, though mining has continued intermittently. See Bessie Doak Haynes, 'Gold on the Queen Charlotte's Island,' *Beaver* 297 (1966):4-11; Margaret Ormsby, 'Introduction,' in *Fort Victoria Letters, 1846-1859*, ed. Hartwell Bowsfield (Winnipeg: Hudson's Bay Record Society 1979), xci-xciv.
231 Moresby Lake.
232 The Newcombe map and Swanton list it as Sqai'tāo, with the comment: 'sometimes spoken of as a town, was only a Haida camp formed during

Notes to pp. 52-7   185

the rush for gold to Gold Harbor.' See Swanton, *Ethnology of the Haida*, 280.

233 The village was Kaisan (Swanton refers to it as Qai'sun) on the west coast of Moresby Island. Skatz-sai gives the name Skotsai Bay, which is behind today's Saunders Island. The chief, according to Swanton, was Nāñ na'gage shilxa'ogas, the last name becoming Dawson's Skatz-sai and others' Skotsai. See Swanton, *Ethnology of the Haida*, 280; and MacDonald, *Haida Monumental Art*, 117.

234 Dawson commented in his published report that 'the *blanket* is now, however, the recognised currency, not only among the Haidas, but generally along the coast. It takes the place of the beaver-skin currency of the interior of British Columbia and the North-west Territory. The blankets used in trade are distinguished by points, or marks on the edge, woven into their texture, the best being four-point, the smallest and poorest one-point. The acknowledged unit of value is a single two-and-a-half-point blanket, now worth a little over $1.50.' See Dawson, 'On the Haida,' 128-9.

235 Chinook Jargon, 'a very big man indeed, big chief.'

236 Torrens Island.

237 In 1884, Newton H. Chittenden estimated a population of 100 at Skidegate and 108 at New Gold Harbour. See Chittenden, *Exploration of the Queen Charlotte Islands*, 36. New Gold Harbour was abandoned in 1893.

238 Probably the Pacific cod, *Gadus macrocephalus* Tilesius.

239 Swanton's Tc!ā'ał on the south side of Chaatl Island. See Swanton, *Ethnology of the Haida*, 280.

240 Swanton's Qai'sun, east of Annesley Point at the entrance to Inskip Channel on the west coast of Moresby Island. See Swanton, *Ethnology of the Haida*, 280. For a description, see MacDonald, *Haida Monumental Art*, 116-20.

241 Sea otter, *Enhydra lutris lutris* (L.).

242 By principle of matrilineal descent, inheritance was by the sister's son.

243 Probably Armentiéres Channel on the east side of Chaatl Island.

244 Trounce Inlet.

245 'The septum of the nose is generally perforated in both males and females, and was formerly made to sustain a pendant of haliotis shell or a silver ring, though it is not now used in this way.' See Dawson, 'On the Haida,' 103.

246 Or '*spose hullel konaway siwash hi-yu he-he*,' Chinook Jargon for 'when shaken, all Indians laugh a lot.'

247 '*Mache*,' is unlisted in Chinook Jargon dictionaries. Fortunately, Dawson provides his own translation.

248 Fife Point.

249 Hiellen River.

250 Dawson puts it a mile and a half west of the Hiellen River and writes of 'several rude houses, inhabited by the Masset Indians during a portion of

the summer while they are engaged in curing halibut and making dogfish oil.' See Dawson, 'Report on the Queen Charlotte Islands,' 34. It is at the site of the former Yagan village, Swanton's Yā'gʌn, in Swanton, *Ethnology of the Haida*, 281.
251 9 August 1878, GSC256-C2 (C 51370); GSC257-C2 (C 51371); GSC258-C2 (C 51369).
252 Yakan Point.
253 10 August 1878, GSC259-C2 (PA 38149).
254 Martin H. Offutt had established the Masset post in 1870 and retired in December 1878, though a Mr. A. Cooper had replaced him from 1871 to 1874. Offutt was not an old company man, apparently joining the service recently, perhaps only for the Masset posting. He had become known to the Masset Haida through his residency among the Tsimshian of Port Simpson. His wife was Tsimshian. See Dalzell, *The Queen Charlotte Islands, 1774-1966*, 76-7.
255 His camp of 6-8 August, a few miles north of Lawn Hill.
256 See George Mercer Dawson, Field Notebooks, RG 45, vol. 292, no. 2803, 76v-77, NAC. The tertiary age of Lawn Hill is confirmed in A. Sutherland Brown, *Geology of the Queen Charlotte Islands*, British Columbia Department of Mines and Petroleum Resources Bulletin, no. 54 (Victoria, BC: Department of Mines and Petroleum Resources 1968), 115 and fig. 5, sheet B.
257 The 'grass' is impossible to identify; 'beach peas,' seashore peavine, *Lathyrus japonicus* Willd.
258 The company lost two vessels named *Vancouver* on the spit: the first in 1834, the second in 1854.
259 'Twenty-one miles in width.' See Dawson, 'Report on the Queen Charlotte Islands,' 34.
260 William Henry Collison (1847-1922), who was born in Ireland and attended the Church Missionary College at Islington, came to the northern British Columbia coast in 1873 to assist William Duncan. Arriving on the Queen Charlotte Islands at Masset in 1876, Collison later worked on the upper Skeena River with the Gitskan and was archdeacon of Metlakatla from 1891 until 1921 when he retired. For Collison's missionary work, see his *In the Wake of the War Canoe* (Toronto: Musson 1916).
261 William Duncan (1832-1918) was a prominent Anglican lay missionary who established the Christian Indian village of Metlakatla in 1862, near the mouth of the Skeena River. Duncan took special care to encourage a variety of industries that would further the Native people's economic circumstances and also drastically alter many of their traditional ways. Unfortunately, Duncan's efforts were later marred by bitter strife, and, in 1887, he led a band of followers to Alaska to set up a new village. See Jean Usher, *William Duncan of Metlakatla: A Victorian Missionary in British Columbia*, National Museum of Man Publications in History, no. 5 (Ottawa:

National Museums of Canada 1974); and Peter Murray, *The Devil and Mr. Duncan: A History of Two Metlakatlas* (Victoria: Sono Nis Press 1985).
262 The celebrated Northwest Coast potlatch varied from group to group. For the Haida potlatch, see Dawson, 'On the Haida,' 119-21; Swanton, *Ethnology of the Haida*, 155-81; George Peter Murdock, *Rank and Potlatch among the Haida* (New Haven: Yale University Press 1935); and Abraham Rosman and Paula G. Rubel, *Feasting with Mine Enemy: Rank and Exchange among Northwest Coast Societies* (New York: Columbia University Press 1971), 34-68.
263 Actually crests, which are owned by families and include a large number of animals, fish, and other natural phenomena. See Swanton, *Ethnology of the Haida*, 107-18.
264 Or clan. The Haida are divided into two exogamous phratries or clans, the Raven and the Eagle.
265 Perhaps Watun River.
266 According to J.W. Dawson, who described these eastern Canadian deposits, 'The Saxicava sand, in typical localities, consists of yellow or brownish quartzose sand, derived probably from the waste of the Potsdam sandstone and Laurentian gneiss, and stratified. It often contains layers of gravel, and sometimes is represented altogether by coarse gravels.' See J. William Dawson, *The Canadian Ice Age: Being Notes on the Pleistocene Geology of Canada* (Montreal: William V. Dawson 1893), 60.
267 There were a number of such camps along Masset Sound. This may, however, be the site of the village of Gîtînq!a. See Swanton, *Ethnology of the Haida*, 281.
268 Canada geese, *Branta canadensis fulva* Delacour.
269 Masset Inlet.
270 Yakoun River.
271 Perhaps Dr. Kwude, a lineage chief and well-known Masset shaman.
272 British or Canadian as opposed to 'Boston men,' Americans.
273 Or '*Okuke illahee King George illahee ahnkuttie Bostons tiky kapswolla pe mika kloshe nanitch*,' Chinook Jargon for: 'This country is English; long ago Americans wanted to steal. Be careful.'
274 '*Hilo*,' Chinook Jargon for 'no.'
275 Juskatla Inlet.
276 Juskatla Narrows.
277 Fraser Island.
278 Mamin village occupied by the Mamum-Gituni of Masset. See Dalzell, *The Queen Charlotte Islands*, vol. 2, 409. It is not recorded in Swanton, *Ethnology of the Haida*.
279 Masset Inlet.
280 Probably McClinton Bay, referred to as Tin-in-ow-a on Dawson's map.
281 Dinan Bay.

282 Ain Lake or the larger, more distant Ian Lake.
283 Dalzell notes that 'an old canoe is said to lie in the vicinity of this lake. Ranging in length from 50 to 90 feet, depending on the narrator, quite finished it was left covered with shakes awaiting help with launching. It is said that the builders succumbed to small-pox before this could be effected.' See Dalzell, *The Queen Charlotte Islands*, vol. 2, 422.
284 Dalzell, *The Queen Charlotte Islands*, vol. 2, 422, calls it Kogals-Kun village while the Newcombe-Swanton map renders it K'ogālskun. See Swanton, *Ethnology of the Haida*, 279.
285 Langara Island.
286 Probably a reference to Juan Pérez on the *Santiago* out of San Blas, Mexico, who arrived off Langara Island on 18 July 1774. On the following morning he was met by three Haida canoes followed by more later in the day. See Warren L. Cook, *Flood Tide of Empire: Spain and the Pacific Northwest, 1543-1819* (New Haven: Yale University Press 1973), 54-65.
287 Ut-te-was, in Dawson's 'On the Haida,' containing 'about twenty houses, counting both large and small,' with over forty carved posts. This was the village called Masset, now Haida. Swanton lists it as āatē'was, in *Ethnology of the Haida*, 281. In 1891, when Dawson visited the Queen Charlotte Islands on his return from the Bering Sea, he 'found Masset considerably changed since I saw it 13 years ago. Still many totem poles standing and some of them in good preservation, but most of the old Native houses have either disappeared or have been abandoned. Commonplace frame houses of rather small size taking their places.' See Dawson's Journal, 2 October 1891, G.M. Dawson Papers, MUA. For a description, see MacDonald, *Haida Monumental Art*, 131-70.
288 Ka-yung, in Dawson's 'On the Haida,' 155-6, where he mentions that it was not particularly examined. Listed as Q!ayā'ñ by Swanton, *Ethnology of the Haida*, 281.
289 Yān, in Dawson, 'On the Haida,' 155, where he describes it as showing 'about twenty houses new and old, with thirty carved posts.' It is Swanton's Yan, *Ethnology of the Haida*, 281.
290 'The length of Masset Sound from its seaward entrance to the point at which it expands widely is nineteen miles.' See Dawson, 'Report on the Queen Charlotte Islands,' 35.
291 Maast Island.
292 Called by the Indians Tsoo-skatli. See Dawson, 'Report on the Queen Charlotte Islands,' 36.
293 Yakoun River.
294 Yakoun Lake.
295 Either Gāli'nskun or Lā'nas. See Swanton, *Ethnology of the Haida*, 280.
296 About seven miles north of Skidegate village.
297 Masset Inlet.

298 Mamin River.
299 See 16 August 1878.
300 Pink salmon, *Oncorhynchus gorbuscha* (Walbaum).
301 The arms of Masset Inlet appear to be glacier-scoured channels formed by ice tongues moving northeasterly from the high country of southwestern Graham Island.
302 The volcanic rocks are indeed Tertiary, with a radiometric date indicating a Miocene age.
303 'Chalcedony' was a cryptocrystalline variety of silica, consisting essentially of fibrous or ultrafine quartz, some opal, together with water, which is either enclosed in the lattice or in the macrostructure of the mineral. 'Agates' were banded chalcedonic silica.
304 'Trachytic' is a textual term applied to volcanic rocks in which feldspar microlites of the groundmass have a subparallel arrangement corresponding to the flow lines of the lava from which they were formed. 'Acidic' is a term applied to any igneous rock composed predominantly of light-coloured mineral having a relatively low specific gravity.
305 Mamin River.
306 The eruptive centres have not been established.
307 For the glacial record of this region see John J. Clague, Rolf W. Mathewes, and Barry G. Warner, 'Later Quaternary Geology of Eastern Graham Island, Queen Charlotte Islands, British Columbia,' *Canadian Journal of Earth Sciences* 19 (1982):1786-95.
308 Probably the green sea urchin, *Strongylocentrotus droebachiensis* (Müller).
309 The 'bladder weed' was bladder wrack or rockweed, *Fucus furcatus* C. Agardh.; and 'Eel Grass' was eel-grass, *Zostera marina* L.
310 18 August 1878, GSC260-C2 (PA 44332) and GSC262-C2 (PA 44327).
311 The right page of the journal covers elevations and plans of a Haida house, one that was almost literally reproduced in Dawson's 'Report on the Queen Charlotte Islands,' plate XIV. In the journal it is labelled: 'Diagram with measurements of Indian home at Virago Sound Q.C.I. Type of many but rather larger than now usual. The upright boarding of the ends & sides, with the roofing, is omitted for greater clearness.'
312 Located at Mary Point on Alexandra Narrows between Virago Sound and Naden Harbour. Also known as Nightasis, Kung was settled by Albert Edward Edenshaw and his followers in 1853. Edenshaw moved to Yats (Yatza) in 1875, but Kung was not completely abandoned until a few years later. It is Swanton's Qañ, *Ethnology of the Haida*, 281. Dawson elaborates that 'it has been a substantial and well-constructed one, [village] but is now rather decayed, though some of the houses are still inhabited. The houses arranged along the edge of a low bank, facing a fine sandy beach, are eight or ten in number, some of them quite large. The carved posts are not very numerous, though in a few instances elaborate.' See Dawson,

'On the Haida,' 155. See also MacDonald, *Haida Monumental Art*, 177-83.
313 Swanton's Yats 1, *Ethnology of the Haida*, 281. Established by Albert Edward Edenshaw in 1875 on the northern coast, east of Virago Sound, to attract trade from the Kaigani Haida of Prince of Wales Island. It is doubtful if it was meant to be an all-year village site, and it was abandoned in 1883. Dawson adds: 'Like many of the Haida villages, its position is much exposed, and it must be difficult to land at it with strong northerly and northeasterly winds. This village site is quite new, having been occupied only a few years. There are at present eight or ten roughly built houses, with few and poorly carved posts. The people who formerly lived at the entrance to Virago Sound are abandoning that place for this, because, as was explained to me by their chief, Edensaw, they can get more trade here, as many Indians come across from the north. The traverse from Cape Kygane or Muzon to Klas-kwun is about forty miles, and there is a rather prominent hill behind the point by which the canoe-men doubtless direct their course. At the time of our visit, in August 1878, a great part of the population of the northern portion of the Queen Charlotte Islands was collected here preparatory to the erection of carved posts and giving away of property, for which the arrival of the Kai-ga-ni Haidas was waited, these people being unable to cross owing to the prevalent fog and rough weather.' See Dawson, 'On the Haida,' 154-5. See also MacDonald, *Haida Monumental Art*, 184.
314 The points along the north coast are confirmed as Tertiary rocks of Masset Formation in Sutherland Brown, *Geology of the Queen Charlotte Islands*, 115 and fig. 5, sheet c.
315 The 'wide funnel shaped mouth' is now called Virago Sound, while the 'land-locked harbour' is Naden Harbour.
316 Alexandra Narrows.
317 Kung village.
318 Naden River.
319 Eden Lake.
320 Kung, putting the camp on the east shore.
321 Probably a basking shark, *Cetorhinurs maximus* (Gunnerus).
322 This would actually put him in Naden Harbour.
323 Yats (Yatza).
324 Towustasin Hill.
325 Dawson was probably referring to the Sitka spruce, *P. sitchensis*, rather than to the Douglas-fir, *P. menziesii* (*'Abies Menziesii'*), since the latter's cambium was not used by the Haida. Also, in his published report Dawson refers to '*A. Menziesii*' as 'the spruce.' See Dawson, 'On the Haida,' 107. According to Nancy Turner, Dawson was probably correct about Haida utilization of mountain hemlock and the non-use of the shore pine, *Pinus contorta* Dougl. ex Loud. See Nancy J. Turner, *Food Plants of British Colum-*

bia Indians, part 1, *Coastal Peoples*, British Columbia Provincial Museum Handbook, no. 34 (Victoria: British Columbia Provincial Museum 1975), 65.

326 I.e., sufficient quantity.
327 The wild crabapple, *Pyrus fusca* Raf., was widely used by all coastal Indians, including the Haida. Dawson's description of Haida methods of use corresponds closely to that of modern ethnobotanists. See Turner, *Food of the Coastal Peoples*, 204.
328 Albert Edward Edenshaw (or Edensaw) (1810?-94) was one of the most prominent Haida of the nineteenth century. He was suspected of complicity in the 1852 plunder of the *Susan Sturgis* off Rose Point Spit. He established Kung-in 1853 and Yats (Yatza) in 1875, before settling in Masset in 1883. Commander James C. Prevost described him as 'decidedly the most advanced Indian I have met with on the Coast: quick, ambitious, crafty and, above all, anxious to obtain the good opinion of the white man.' See Barry M. Gough, 'New Light on Haida Chiefship: The Case of Edenshaw, 1850-1853,' *Ethnohistory* 29 (1982):131-9; Mary Lee Stearns, 'Succession to Chiefship in Haida Society,' in *The Tsimshian and Their Neighbors of the North Pacific Coast*, ed. Jay Miller and Carol M. Eastman (Seattle: University of Washington Press 1984), 206-17; and 'Eda'nsa,' *Dictionary of Canadian Biography* 12:289-91.
329 These are the Kaigani Haida who migrated in the eighteenth century to the southern portion of Prince of Wales Island.
330 Dixon Entrance.
331 The American elk or wapiti, *Cervus canadensis nelsoni* Bailey, was not native to the Queen Charlotte Islands but has been introduced to Graham Island. Another subspecies, *Cervus canadensis roosevelti* Merriam, is native to Vancouver Island.
332 The 'small fish with bright red flesh' was the sockeye salmon, *Oncorhynchus nerka* (Walbaum) and 'silver Salmon,' probably the pink salmon, *O. gorbuscha*, even though 'silver salmon' is a variant vernacular for coho salmon, *Oncorhynchus kisutch* (Walbaum).
333 The 'trout' were probably rainbow trout, *Salmo gairdneri* Richardson, or possibly coastal cutthroat trout, *Salmo clarki clarki* Richardson, or the Dolly Varden, *Salvelinus malma* (Walbaum).
334 Yats or Yatsa. See 19 August 1878.
335 Kung. See 19 August 1878.
336 Klashwun or Little Mountain, about 340 feet on its west side and visible for a considerable distance. See Dalzell, *The Queen Charlotte Islands*, vol. 2, 447.
337 Chief Weah.
338 The incident occurred on 25 September 1852. Edenshaw, acting as a guide, was aboard the *Susan Sturgis* with his wife and child when it was attacked

by a group of Masset Indians just off Rose Point. According to the ship's captain, Matthew Rooney, Edenshaw tried to dissuade the attackers but Chief Weah of the Massets refused to listen. Edenshaw, seeing the futility of his attempts, then joined in the plunder. He did, however, manage to bargain for the lives of the captain and his crew. Edenshaw was suspected of arranging the entire affair, but his complicity was never proved. See Gough, 'New Light on Haida Chiefship,' 131-9.

339 23 August 1878, GSC265-C2 (PA 38154) and GSC266-C2 (PA 38147). The latter print is reproduced and discussed in Margaret B. Blackman, '"Copying People": Northwest Coast Native Response to Early Photography,' *BC Studies* 52 (1981-2):92 and plate II.

340 Dadens.

341 Hiellen, Swanton's Łi'elañ (*Ethnology of the Haida*, 280), on the east bank of the Hiellen River near Rose Spit, was Edensaw's original home. In the early 1840s he led his Sta'stas lineage to K!iŭ'sta and reasserted dominance over the K'awas lineage, who had already settled there. For Edensaw's moves, see Mary Lee Stearns, 'Succession to Chiefship,' 204-18.

342 Klashwun Point.

343 Jalun River, named by Dawson.

344 William Douglas was commander of the British merchant ship the *Iphigenia Nubiana*. He was off Langara Island in August 1788 and returned for a week's stay in June 1789. The reference to an earlier visitor was perhaps to Pérez.

345 Or pelagic cormorants, *Phalacrocorax pelagicus pelagicus* Pallas.

346 Pillar Rock, so named by Dawson, in Pillar Bay. 24 August 1878, GSC267-C2 (PA 51118). The view was photo-lithographed and included in Dawson, 'Report on the Queen Charlotte Islands,' plate II, opp. 40.

347 Dadens.

348 McPherson Point.

349 Cloak Bay and Cox Island.

350 Pillar Rock, at 24 August 1878.

351 K!iŭ'sta. See Swanton, *Ethnology of the Haida*, 281.

352 Lepas Bay, named by Dawson.

353 Probably Lauder Point.

354 Parry Passage.

355 Perhaps the spiny sun star, *Crossaster papposus* Linnaeus.

356 Gnarled Islands.

357 Probably Mount Griffin.

358 West southwest of Jacinto Point, Zayas Island.

359 Caamaño Passage from Dixon Entrance.

360 30 August 1878, GSC269-C2 (PA 44333) and GSC270-C2 (PA 44344).

361 This is Joseph William McKay of the Hudson's Bay Company. The

'Chimseyan Ledge,' on the hill behind Port Simpson, was, according to Dawson, composed of 'pyrrhotite with copper-pyrites in quartz tongue.' See Dawson, 'The Mineral Wealth of B.C.,' 153.

362 A fort was first built in 1831-2 near the mouth of the Nass River but was moved to a location on the Tsimpsean Peninsula in 1834. In 1860, a new fort was built on the same site. For more on Fort Simpson, see Daniel Clayton, 'Geographies of the Lower Skeena,' *BC Studies* 94 (1992):29-40.

363 Thomas Crosby (1840-1914), a prominent Methodist missionary to the Tsimshian, came to Fort (Port) Simpson in 1874. For over twenty years, Crosby zealously furthered Methodist missions, establishing a number of outlying stations and initiating a marine mission as well as developing Port Simpson after the model pioneered by William Duncan. For accounts of his work, see Thomas Crosby, *Up and Down the North Pacific Coast by Canoe and Mission Ship* (Toronto: Missionary Society of the Methodist Church 1914); Clarence R. Bolt, 'The Conversion of the Port Simpson Tsimshian: Indian Control or Missionary Manipulation?,' *BC Studies* 57 (1983):38-56; and Clarence R. Bolt, *Thomas Crosby and the Tsimshian: Small Shoes for Feet Too Large* (Vancouver: UBC Press 1992).

364 Possibly killdeer, *Charadrius vociferus* L.

365 2 September 1878, GSC271-C2 (PA 44335).

366 Robert Tomlinson (1842-1913) was an Irish physician who came to British Columbia in 1867 to assist William Duncan. Tomlinson's first posting was to the Nass River, where he founded a mission station at Kincolith. A staunch supporter of Duncan, he lost the Anglican Church's backing when Duncan left the Church Missionary Society. In 1888, Tomlinson founded his own non-denominational mission, Minskinisht or Cedarvale, on the Skeena River. See also Margaret Whitehead, *Now You Are My Brother: Missionaries in British Columbia*, Sound Heritage Series, no. 34 (Victoria: Provincial Archives of British Columbia 1981), 6-25.

367 Captain William F. Madden, 'ship master' of Port Essington.

368 Unidentifiable. William A. Ashe was one of four sub-assistant astronomers.

369 On the Tsimpsean Peninsula at the mouth of the Skeena River where, in the 1860s, William H. Woodcock had established a public inn for the accommodation of miners going to and from the Omineca country via the Skeena River. The site was known as Woodcock's Landing until, with the 1876 establishment of a salmon cannery there, it gradually became known as Inverness, from Inverness Passage.

370 Chatham Sound.

371 The '*Rhynchinellas*' were the brachipod, *Hemithyris psittacea* (Gmelin); the '*Ledas*,' one of the nut clams, *Nuculana* spp., such as Müller's nut clam, *Nuculana pernula* (Müller), or minute nut clam, *Nuculana minuta* (Fabricius); and the 'brittle star,' the brittle star, *O. lutkeni*.

372 East of Shattock Hill in the entrance to Big Bay.
373 Portland Inlet. The watercolour, dated 29 August 1878, is now in the McCord Museum.
374 Dawson's information in the following was partly reliable. The Tsimshian were not closely related to the Dene nor are their languages related. Tsimshian early history is obscure and controversial, but any migration occurred earlier than the century mentioned by Dawson. Their name is, however, derived from their word for the Skeena, *K-shian* or *ikshean*, river of mists, and thus 'people of the river of mists.'
375 Robert H. Hall was a Hudson's Bay Company clerk at Fort Simpson, who had previously served at Stuart Lake.
376 Stikine River. The Tongas are a Tlingit group.
377 For a description of Haida puberty rites, see Swanton, *Ethnology of the Haida*, 48-50.
378 Stone-boring clams of the Lamellibranchiate family.
379 Work Channel.
380 The deer family. Their distribution has changed considerably since Dawson received his information.
381 British Columbia moose, *Alces alces andersoni* Peterson. See David J. Spalding, 'The Early History of Moose (*Alces alces*): Distribution and Relative Abundance in British Columbia,' Royal British Columbia Museum, *Contributions to Natural Science* 11 (1990):1-12.
382 Mountain caribou, *Rangifer tarandus montanus* Seton.
383 It is difficult to know what deer Dawson was describing, since the only species found in the range outlined was the mule deer, *Odocoileus hemionus hemionus* (Rafinesque), a large deer.
384 The grizzly bear, *Ursus arctos horribilis* Ord. 'Recent studies have indicated that all these supposed species [of the grizzly bear] refer to individual variations in one species of bear.' See A.W.F. Banfield, *The Mammals of Canada* (Toronto: University of Toronto Press 1974), 311.
385 A copper ore mineral.
386 Three months earlier, *The Colonist* had carried a report of Captain Madden's interest in the Skeena quartz ledge, 'prospected four years ago' and 'believed to be very rich.' See *Victoria Daily British Colonist*, 9 June 1878.
387 7 September 1878, GSC272-C2 (PA 152369).
388 Douglas Channel.
389 Probably 'Mt. Gill & Whale Passage from Wright Sd.,' a watercolour, dated 7 September 1878, in the McCord Museum.
390 The waterfall flows into Marmot Cave.
391 The lake is ungazetted.
392 The stream is ungazetted.
393 9 September 1878, GSC273-C2 (PA 152368).
394 Now the Kokyet reserve, on the extreme southwest coast of Yeo Island.

The Kokyet or Kokyitoch are Heiltsuk speakers, closely related to the Bella Bella Heiltsuk speakers.
395 Spiller Channel.
396 Probably Thomas Laughton, postmaster for the Hudson's Bay Company at Bella Bella from 1871 to 1873. An ex-Royal Navy man, Laughton had been a trader at Port San Juan in the 1860s.
397 Dawson's reference is to his report on explorations conducted in the 1877 field season in south-central British Columbia. His 'Preliminary Report on the Physical and Geological Features of the Southern Portion of the Interior of British Columbia, 1877,' Geological Survey of Canada, *Report of Progress for 1877-78* (Montreal: Dawson Brothers 1879), Report B, was not published until 1879.
398 Flood myths are common among British Columbia Native people.
399 This Heiltsuk story appears, without the same ending, in Franz Boas, *Bella Bella Texts* (New York: Columbia University Press 1928), 36-45.
400 Aboriginal populations were declining significantly during the nineteenth century. Introduced diseases were the major cause. For a discussion of the issue, see Wilson Duff, *The Impact of the White Man*, vol. 1, *The Indian History of British Columbia*, Anthropology in British Columbia Memoir, no. 5 (Victoria: British Columbia Provincial Museum 1969), 40-4; and Robert S. Boyd, 'Demographic History, 1774-1874,' in *Handbook of North American Indians*, ed., William G. Sturtevant, vol. 7, *Northwest Coast* (Washington: Smithsonian Institute Press 1991), 135-48.
401 Or '*chako memaloose*,' Chinook Jargon for 'have died.'
402 Chinook Jargon for 'I don't know what God is doing.'
403 Nawitti Kwakiutl from Hope Island and Lekwiltok Kwakiutl from Cape Mudge. Dawson had encountered the Nawitti already, 4 June 1878, and would again, 14-15 September 1878.
404 A reference to his survey publication 'On the Haida.'
405 The whales are impossible to identify precisely.
406 Cape Sutil.
407 See Boas, 'The Houses of the Kwakiutl,' for an engraving of Cheap's house and pole in 1886.
408 The 'holithurians' were sea cucumbers of the class Holothuroidea; 'sea eggs,' sea urchins such as the green sea urchin, *S. droebachiensis*; and 'star fish,' one of several species such as the common starfish, *P. ochraceus*.
409 Dawson is quoting the first lines of an untitled Tennyson poem. For the full text, see Alfred Lord Tennyson, *The Poems of Tennyson*, ed. Christopher Ricks (London: Longmans, Green 1969), 602.
410 Robert Hunt (1828-93) came to Victoria from Dorset in the spring of 1850. He worked for the Hudson's Bay Company at Fort Rupert, and, when the company closed their post in 1885, he acquired the property. Hunt remained at Fort Rupert until his death. He was married to a Mary Ebbetts,

a Tongas Tlingit. Among his several children was George Hunt (1854-1933), collaborator with Franz Boas on Kwakiutl ethnology, folklore, and artifact collecting.

411 For Dawson's discussion of these tribes and a published list of tribes and localities see his 'Notes and Observations on the Kwakiool People of the Northern Part of Vancouver Island and Adjacent Coasts, Made during the Summer of 1885, With a Vocabulary of about Seven Hundred Words,' *Proceedings and Transactions of the Royal Society of Canada* 5 (1887), sec. II, 65-75.

412 Quadra Island.

413 Alfred James Hall (1853?-1918) was ordained in England in 1877 and shortly thereafter came to Metlakatla under the auspices of the Church Missionary Society. In 1878, he moved to Fort Rupert where he remained for two years before moving the mission to Alert Bay in 1880. For some thirty years Hall laboured there, establishing a residential school and a sawmill. Hall also mastered the Native language and in 1888 published his 'A Grammar of the Kwakiutl Language,' *Proceedings and Transactions of the Royal Society of Canada* 6 (1888), sec. II, 59-106. See also Barry M. Gough, 'A Priest versus the Potlatch: The Reverend Alfred James Hall and the Fort Rupert Kwakiutl, 1878-1880,' *Journal of the Canadian Church Historical Society* 24 (1982):75-89.

414 Quatse River.

415 2 October 1878, GSC275-C2 (PA 51117) and GSC276-C2 (PA 44336).

416 Rankine arrived in Montreal only the next day, 3 October 1878.

417 Dawson was talking about the Suquash coal fields, which were first worked by the Hudson's Bay Company in 1836. Later, in the early 1850s, these deposits were abandoned when the richer finds around Nanaimo became known. See Dawson's description of the Suquash coal in his 'Report on Geological Examination of the Northern Part of Vancouver Island and Adjacent Coasts,' in Geological Survey of Canada, *Annual Report, 1886*, n.s., 2 (1887):62-70, Report B.

418 See 13 October 1878.

419 Dawson's report for the 1876 season, 'Report on Explorations,' which was printed in 1878.

420 Alfred Richard Cecil Selwyn (1824-1902) was director of the Geological Survey of Canada from 1869 to 1895, when Dawson assumed the position upon Selwyn's retirement. It was under Selwyn's direction that Dawson and his contemporaries began the immense task of surveying the western portion of British North America. For biographical details, see H.M. Ami, 'Sketch of the Life and Work of the Late Dr. A.R.C. Selwyn,' *American Geologist* 31 (1903):1-21.

421 Dawson's brother, William Bell, had just returned from completing his course at Paris' Ecole des Ponts et Chaussées. The essay is unidentifiable.

He was now seeking a position at the University of Toronto for which he was not successful.
422 Port McNeill.
423 A Montreal newspaper.
424 See 15 September 1878.
425 Dawson further elaborates: 'The houses are generally large, and are used as dwelling places by two or more families, each occupying a corner, which is closed in by temporary partitions of split cedar planks, six or eight feet in height, or by a screen of cloth on one or two sides. Each family has, as a rule, its own fire, with cedar planks laid down near it to sit and sleep on ... The household effects and property of the inmates are piled up round the walls, or stowed away in little cupboard-like partitioned spaces at the sides or back of the house. Above the fire belonging to each family is generally a frame of poles or slips of cedar, upon which clothes may be hung to dry, and dried fish or dried clams are stored in the smoke.' See Dawson, 'Notes and Observations on the Kwakiool People,' 75.
426 The 'commonest form' of wooden dish was a canoe-shaped feast dish, indeed a common form. See Audrey Hawthorne, *Kwakiutl Art* (Seattle: University of Washington Press 1979), 182, fig. 335. The dish with 'indians clasping the vessel' was a bowl grasped by two human figures, also a fairly common form. The 'large spoon' is more unusual. It is a ladle bearing a frog caught in the mouth of a bird, usually thought to be a kingfisher, and is a motif most commonly associated with the raven rattle.
427 Vere Cove.
428 Walkem Islands.
429 Seymour Narrows.
430 Discovery Passage.
431 Arbutus, *Arbutus menziesii* Pursh.
432 Gilbert Malcolm Sproat (1834-1913) was born in Scotland and came to Vancouver Island in 1860 under the employ of Anderson & Company, operators of a sawmill near Alberni. After returning to London, where he was agent-general for British Columbia, Sproat came back to British Columbia and was appointed joint commissioner on the Indian Land Commission in 1876. Upon resigning from that position in 1880, Sproat spent much time in the Kootenay region in various capacities such as gold commissioner and assistant commissioner of lands and works. See T.A. Rickard, 'Gilbert Malcolm Sproat,' *British Columbia Historical Quarterly* 1 (1937):21-32.
433 An M. Watt is listed in the *Guide to the Province of British Columbia for 1877-8* (Victoria: T.N. Hibben & Co. 1877). The workings of the Baynes Sound Coal Company were located on the Tsable River, near Fanny Bay, almost three miles due west from the river's mouth. Organized in October 1875 and working in April 1876, the colliery employed fifty-five men during

1876. By 1 November 1876, some six hundred tons of coal had been produced, but the mine's lifespan was short, and, with a falling market for coal in 1877, the company ceased production.
434 Only a single photograph remains. 14 October 1878, GSC278-C2 (PA 51107).
435 Such middens were fairly common along coastal areas.
436 John Bryden (1831-1915) was mines manager for the Vancouver Coal Mining and Land Company. Bryden disagreed with Mark Bate's lenient labour policies and management practices and later left the company to work for his father-in-law, Robert Dunsmuir. Rising to the position of managing partner in the Dunsmuir organization, Bryden became one of the wealthiest men in British Columbia.
437 The New Douglas or Chase River Mine, operated from 1875 to 1886 by the Vancouver Coal Mining and Land Company.
438 John J. Landale (1836?-86) was a well-known civil and mining engineer. See his obituary in the *Nanaimo Free Press*, 9 January 1886.
439 William Sutton arrived in British Columbia in 1875 from Ontario, where he had been sheriff of Bruce County. With his three sons, Sutton built a sawmill in 1877 and carried on a considerable logging operation. He was granted a timber lease of 7,069 acres in the Lake Cowichan area in 1879. His enterprises did not prove successful, however, and he later sold the operations.
440 Actually at Genoa Bay on the north side of Cowichan Bay.
441 The 'old workings' were the so-called Old Douglas Mine near the centre of the settlement of Nanaimo, first mined in 1852. For a map illustrating the Douglas seam and locations of the two workings, see J.E. Muller and M.E. Atchison, *Geology, History, and Potential of Vancouver Island Coal Deposits*, Geological Survey of Canada Paper, no.70-53 (Ottawa: Information Canada 1971), fig. 12. For a concise outline of the geology of the Douglas seam, see Charles H. Clapp, *Geology of the Nanaimo Map Area*, Geological Survey of Canada Memoir, no. 51 (Ottawa: Government Printing Bureau 1914), 110-14.
442 Wallace Island.
443 John Hamilton Gray (1814-89) had been a prominent New Brunswick lawyer and politician before coming to British Columbia. A strong advocate of Confederation, Gray represented New Brunswick at the 1864 Charlottetown Conference and served in the House of Commons from 1867 to 1872. Then, realizing he had only a limited political future, Gray sought and obtained a puisne judgeship on the British Columbia Supreme Court in 1872. See 'Gray, John Hamilton,' *Dictionary of Canadian Biography*, 11:372-6.
444 Ernest Barron Chandler Hanington (1851-1916) graduated from McGill Medical School in 1875 and acted as medical superintendent at the general hospital in Saint John, New Brunswick, before coming to British Co-

lumbia in 1878. On his arrival in the province, Hanington became medical officer to the labourers building the Canadian Pacific Railway in the lower Fraser River. He advanced to chief surgeon at Yale for the railway contractors before leaving in 1885 to enter private practice in Victoria. He had married Ida Tilley Peters of Saint John in 1878. See his obituary in the *Victoria Times*, 11 May 1916.

445 A reference to the firm of Muirhead & Mann, a Victoria contracting and carpentry business.

446 Sir Henry Pering Pellew Crease (1823-1905) and his wife Sarah (Lindley) Crease (1826-1922) were prominent Victorians with whom Dawson spent much time in the winter of 1875-6. Crease came to British Columbia in 1858 and was joined by his wife and children in 1860. After being called to the bar in 1859, he became attorney general of British Columbia, a position he held from 1861 to 1866. In 1870 he was made a puisne judge. Crease played an important role in British Columbia's entry into Confederation, drafting the Terms of Union of 1870 and opening debate on the subject of Confederation. See Margaret A. Ormsby, ed., *A Pioneer Gentlewoman in British Columbia: The Recollections of Susan Allison* (Vancouver: UBC Press 1976), 143-4.

447 Dr. William Fraser Tolmie (1812-86) was one of the pioneer citizens of Victoria. Tolmie joined the Hudson's Bay Company in 1832 and was assigned to Fort Vancouver on the Columbia River, arriving in 1833. In 1859, he came to Victoria to manage the farms of the Puget's Sound Agricultural Company. Tolmie was also a member of the Legislative Assembly of Vancouver Island and later the representative for Victoria in the provincial legislature. Upon his retirement in 1870, Tolmie spent much time on his own eleven-hundred acre Cloverdale farm. See 'Tolmie, William Fraser,' *Dictionary of Canadian Biography*, 11:855-8.

448 Later published as W. Fraser Tolmie and George M. Dawson, *Comparative Vocabularies of the Indian Tribes of British Columbia, with a Map Illustrating Distribution* (Montreal: Dawson Brothers 1884).

449 J.W. Dawson, 'Note on the Phosphates of the Laurentian and Cambrian Rocks of Canada,' *Quarterly Journal of the Geological Society of London* 32 (1876):285-91.

450 Edgar Fawcett (1847-1923), a pioneer who came to Victoria in 1859 and was employed by the customs office for many years. Fawcett wrote extensively about early life in Victoria and British Columbia and was widely acquainted with the region's development. See his *Some Reminiscences of Old Victoria* (Toronto: William Briggs 1912).

451 James Deans (1827-1905) had been assistant to James Richardson in his explorations of the Queen Charlotte Islands. Deans also explored for the Queen Charlotte Coal Company and became known as an ethnologist. In 1892, Deans organized a Haida exhibit at the Chicago World's Fair.

452 'The copper deposit which has received most notice, is situated near the head of Salmon Arm of Jarvis Inlet and between that inlet and Howe Sound. This is owned by the Howe Copper Mining Company. The ore is chiefly Bornite or purple copper and the deposit is not far from the coast, but at an elevation of 3,000 feet above sea-level. It was discovered about 1874, and was worked at intervals between the years 1877-83.' See Dawson, 'The Mineral Wealth of B.C.,' 102.

453 Francis James Roscoe (1831-78), who came to British Columbia in 1862 and settled in Victoria, where he operated a successful hardware business. In 1873 he was elected to the House of Commons, but declined renomination for the 1878 election. See his obituary in the *Victoria Daily British Colonist*, 21 December 1878.

454 Roderick Finlayson (1818-92), a pioneer Hudson's Bay Company man who came to the Pacific coast in 1839. In 1843, Finlayson supervised the construction of Fort Victoria, commanding the post until James Douglas took over in 1849. Appointed a chief factor in 1859, he supervised the Hudson's Bay Company operations in the British Columbia interior from 1862 until his retirement in 1872. Finlayson also served on the Vancouver Island Legislative Council from 1851 to 1863 and was mayor of Victoria in 1878. See 'Finlayson, Roderick,' *Dictionary of Canadian Biography* 12:317-18.

455 Charles Henry Frederick Heisterman (1832-96) owned one of the largest and most profitable real estate businesses in British Columbia. Born in Bremen, Germany, Heisterman settled in England for some years before coming to British Columbia in 1862 and opening his business in 1864.

456 Joseph Hunter (1842-1935) became one of the most recognized surveyors and engineers in British Columbia. Hunter had come to the Cariboo in 1864 as a miner and surveyor and was elected to the provincial legislature. In 1872, he joined the Canadian Pacific Railway Surveys and conducted some of their most memorable expeditions, including a pioneer trek through the Pine River Pass in 1877. In 1883, he was the chief engineer-in-charge of building the Esquimalt and Nanaimo Railway, then general superintendent and vice-president of the line until retiring in 1918. See his obituary in *Victoria Daily Colonist*, 9 April 1935.

457 John Robson (1824-92) played a prominent role in the life of nineteenth-century British Columbia. From 1875 to 1879 he served as paymaster and purveyor to the Canadian Pacific Railway Surveys. He founded the *British Columbian* newspaper in 1861, served in the provincial legislature and ministry, and in 1889 became premier. See 'Robson, John,' *Dictionary of Canadian Biography* 12:914-19.

458 Jacob Duck was an Englishman who came to British Columbia in 1863 to participate in the Cariboo gold rush but instead headed south to what is now Monte Creek, on the South Thompson River. Duck pre-empted land bordering the river and, with Alex Pringle, operated a well-known roadhouse.

459 Now Westwold, southeast of Monte Lake.
460 Thomas Henry Huxley (1825-95) was an eminent British natural scientist who had been one of Dawson's teachers at the Royal School of Mines. Best known as 'Darwin's Bulldog,' he was a renowned scholar and teacher, who wrote extensively in a wide variety of subjects from palaeontology to philosophy, lectured widely, and received numerous awards and distinctions. See 'Huxley, Thomas Henry,' *Dictionary of National Biography, Supplement.*
461 'The Pacific Province (From an Occasional Correspondent of the Witness)' appeared in the *Montreal Daily Witness*, 7 October 1878. Dated on Queen Charlotte Island, 30 July, it was signed 'D.' The Victoria paper which reprinted it was probably the *Evening Standard*, of which no issues for October 1878 are extant.
462 Unidentifiable.
463 Either Arthur or Alfred Fellows, who were partners in the hardware firm of Fellows and Roscoe in Victoria. Later in 1883 the brothers sold their interests to E.G. Prior and spent much time travelling between Victoria and England.
464 Thomas J. Burnes (1832-1915) was a pioneer Victorian known throughout the community. Burnes came to the city in 1858 and for many years operated hotels in Victoria. He later left private business and worked in the customs office until his death. See his obituary in the *Victoria Daily Colonist*, 21 December 1915.
465 Unidentifiable.
466 Captain D.E. Sturt (d. 1885) was a former British Army officer who had been stationed on the San Juan Islands during the dispute in the 1850s and 1860s with the United States over their ownership.
467 Two marble quarries were eventually opened, one near Vananda at the north end of the island and another at Anderson Bay near the southern tip. Neither proved viable operations and only limited development took place. See R.G. McConnell, *Texada Island, B.C.*, Geological Survey of Canada Memoir, no. 58 (Ottawa: Government Printing Bureau 1914), 96-7.
468 Henry Press Wright (1816-92) was a prominent Anglican clergyman who had been appointed the first archdeacon of the Diocese of Columbia in 1861. Wright resigned his post in 1880 and left North America after continued tension with Bishop Hills. See Donald H. Simpson, 'Henry Press Wright: First Archdeacon of Columbia,' *British Columbia Historical Quarterly* 19 (1955):123-86.
469 Amos Bowman (1839-94) began his career as a journalist before spending time with the California Geological Survey. After working with Dawson in 1876, Bowman surveyed and mapped the Cariboo gold fields under the Geological Survey of Canada auspices. He later possessed considerable land holdings in the upper Fraser Valley.

202   Notes to pp. 96-8

470 John Ash (1821-86) obtained his medical training at Guy's Hospital in London, then practised in England for some years before immigrating to Victoria in 1862. In 1865, Ash was elected to the Vancouver Island House of Assembly and, after federation with the mainland, was member for Comox until 1882. From 1872 to 1876, Ash was provincial secretary; in 1874 he was appointed the first British Columbia minister of mines. See 'Ash, John,' *Dictionary of Canadian Biography*, 11:32-3.

471 Stephen Allen Spencer (d. 1911) came to Victoria in 1858 and was in business for some years as a 'daguerreian artist' before disappearing from view for several years then re-emerging in Barkerville in 1871. Spencer subsequently returned to Victoria in 1872 where he operated a number of photographic establishments. Later in the 1880s, he entered into the cannery business at Alert Bay in partnership with West Huson and Thomas Earle. See his obituary in the *Victoria Daily Colonist*, 16 August 1911. See also David Mattison, *The British Columbia Photographers Directory, 1858-1900* (Victoria: Camera Workers Press 1985), s8-9.

472 Possibly William Curtis Ward (d. 1922), an eminent member of Victoria's business elite. The manager of the Bank of British Columbia, Ward had been in Victoria since 1864, when he emigrated from England.

473 Unidentifiable.

474 T.N. Hibben & Company was a respected Victoria publisher, bookseller, and stationery firm, founded by Thomas Napier Hibben in 1858.

475 William Gile Bowman, husband of Sarah A. Bowman, with whom Dawson roomed in the winter of 1875-6, operated a livery stable.

ON THE HAIDA INDIANS OF THE QUEEN CHARLOTTE ISLANDS

Note: These are Dawson's footnotes.

1 The indefinite character of the pronunciation of an unwritten language is so marked, in most of those with which I have had to do, that in the absence of personal familiarity with the language, the use of a complete and highly elaborated system of orthography is in practice almost impossible. I have therefore employed with little alteration, that suggested in No. 160 of the Smithsonian Miscellaneous Collections, entitled *Instructions for Research relative to the Ethnology and Philology of America*. The value of the principal characters used, according to the scheme adopted, is as follows:

a̱ as long in *father*, short in German *hat* (nearly as in English *what*)
e̱ as long in *they*, short in *met*.
i̱ as long in *marine*, short in *pin*.
o as long in *note*, short in *home* or French *mot*.
oo as long in *fool*, *pool*.
u as in *but*.

ai as in *aisle*.
oi as in *oil*.
ow as in *how*.
eu as in *plume*.
y as in *you*.
x represents the gutteral sound sometimes indicated by ch or gh. The long value of vowels is distinguished by the macron, thus ā, ē; the short value by the breve, ă, ĕ.

2 *A Voyage Round the World, but more particularly to the North-west Coast of America. Performed in 1785, 1786, 1787 and 1788, in the King George and Queen Charlotte, Captains Portlock and Dixon.* London, 1789.

3 Meaning simply Indian pig. *Si-wash* from French *sauvage*. *Co-sho* from *cochon*.

4 In another form of the story, it is said that *Ne-kil-stlas* by impregnating two live cockles, and keeping them warm, hatched out both a man and a woman, who were the progenitors of the human race.

5 As sometimes related, it is taken for granted that the sun always was, the moon alone being wanting.

6 Probably refers to the Douglas fir, which here finds its northern limit on the coast, and is very often blackened by fires from the underbrush running up the thick, dry bark of its trunk.

7 Op. cit., p. 198.

8 Cloak Bay and entrance to Parry Passage.

9 Beavers do not occur in the Queen Charlotte Islands, but this term appears to be used here, as elsewhere in the narrative, for sea otter cloaks. See p. 228, in statement on which it is implied that no beaver skins were obtained.

10 Appears to be a species of adze or chisel, as on p. 244, in connection with another part of the N.W. coast, a 'toe made of jasper the same as those used by the New Zealanders,' is mentioned.

11 Now called Masset.

12 In Bruin Bay.

13 Parry Passage.

14 Lucy Island of the chart.

15 Henslung, or the cove to the east of it.

16 A conjecture probably incorrect, for as we have seen, these people were stripped of skins two years before by Dixon, and yet appear to have accumulated a considerable number at the time of Douglas' visit. The ground may have been prepared for the cultivation of the Indian tobacco, referred to on a former page.

17 Possibly to those of Cumshewa Inlet. His latitudes for the southern part of the islands are inexact, as Vancouver remarks.

18 This man may have been the Skidegate chief, and was probably only on a

visit when seen on the west coast. He had no skins to sell at that time.
19 Mr. J.G. Swan incidentally refers to it as *Koona*, p. 5, op. cit.
20 United States Geological and Geographical Survey of the Rocky Mountain Region; Contributions to North American Ethnology. Vol. 1, p. 40.

Index

Ahts. *See* Indians, Nootka
Ain Lake, 62
Ain River, 62, 64, 156
Alert Bay, 17-18, 22
Alexandra Narrows, 66 n. 316
Alliford Bay, 44, 48, 51
Amentières Channel, 54 n. 243
Anderson, Alexander Caulfield, 13
Anthony Island, 25, 161
Arthur Passage, 78
Ash, Dr. John, 95

Bag Harbour, 30 n. 107
Baynes Sound Coal Mine, 87, 90-1, 93
Beaver Harbour, 17, 85
Bella Bella, 20-3, 81
Bent Tree Point, 35 n. 139
Blakow-Coneehaw, Chief, 150-1, 154
Boat Cove, 40 n. 177
Bodega y Quadra, Juan Francisco de la, 153
Bolkus Island, 29, 30
Boston, Chief, 21
Bowman, Amos, 95

Bowman, Sarah A., 12
Bowman, W.G., 95
Bruin Bay, 150 n. 12
Bryden, John, 91
Buck Channel, 51 n. 223
Bull Harbour, 85
Burnaby Island, 4, 30, 32
Burnaby Strait, 29, 30
Burnes, Thomas J., 95

Caamaño, Jacinto, 156
Caamaño Passage, 74 n. 359
Callicum, Chief, 103
Calvert Island, 19
Cape Ball, 57, 64
Cape Calvert, 19
Cape Caution, 19, 83
Cape Fanny, 25
Cape Flattery, 95
Cape Knox, 72; to Port Simpson, 73
Cape Lazo, 15
Cape Mudge, 15-16, 89
Cape Muzon, 155
Cape St. James, 24; to Skidegate Inlet, 24-44, 47

Cape Scott, 85
Cape Sutil, 83 n. 406, 84
Carpenter Bay, 26 n. 97
Cartwright Sound, 51
Chaatl, 43, 160
Chalmers Anchorage, 76
Charles, William, 12
Chatham Sound, 76
Cheap, Chief, 18, 83, 84
Chip, Chief. *See* Cheap, Chief
Cloak Bay, 71 n. 349, 147 n. 8, 158, 162, 163
Clue. *See* Klue
Collins, Mr., 45
Collison, Rev. W.H., 4, 58-9, 66, 68, 95, 97, 115, 164
Copper Bay, 44 n. 200
Copper Islands, 27 n. 102, 32 n. 117
Cordilleran ice sheet, 3
Cormorant Island, 17
Cowgitz coal mine (Queen Charlotte Coal Company), 50, 55, 93
Cowichan, 91
Cox Island, 71 n. 349
Crease, Henry (Sir), 92
Cumshewa, 40, 41-2, 44, 108, 160, 161
Cumshewa, Chief, 41-2
Cumshewa Inlet, 35 n. 142, 40-4, 475-7; to Skidegate Inlet, 44-5
Cumshewa Rocks, 43 n. 195

Dadans, 4, 70 n. 340, 71 n. 347, 151, 152, 154
Dakota, 9, 95
Dall, W.H., 164
Dana Inlet, 37 n. 162
Danger rocks. *See* Garcin rocks
Darwin Sound, 35 n. 137
Dawson, Anna. *See* Harrington, Anna
Dawson, J.W., 23; letters to, 15, 47-8, 87-8, 92-3

Dawson, Margaret, letters to, 9-10, 21-3, 74-5, 81-2, 94
Dawson, Rankine, 9, 10, 12, 22, 23, 24, 48, 50, 61, 74, 75, 81, 93, 94
Dawson, William Bell, 23, 74, 75, 88, 93, 94
Day Point, 23
De la Beche Inlet, 38 n. 143
Deans, James, 93
Denman Island, 90
Departure Bay, 15
Devils Ridge, 73
Digby Island, 78
Dinan Bay, 61 n. 281
Discovery Passage, 89 n. 439
Dixon, George, 98, 102, 112, 147, 153, 158, 160, 161, 162-3
Dixon Entrance, 68, 74
Dodd Narrows, 91
Dogfish oil, 14, 47
Dolomite Narrows, 30
Douglas, Captain Abel, 9, 11, 13, 149-52, 154
Douglas, William, 70
Douglas Channel, 79
Douglas Inlet, 52
Driard House Hotel, 9, 92
Duck, Jacob, 94
Duck and Pringle, 94
Duncan, William, 58, 75, 76, 95, 117
Dundas Island, 73
Dupont, Clara Elizabeth, 12
'Dutch Charley,' 22, 24, 28, 50, 88, 91, 92

East Narrows (Skidegate Channel), 51, 54
East Thurlow Island, 16 n. 44
Echo Harbour, 35
Eden Lake, 66 n. 319
Edenshaw, Chief Albert Edward, 67-8, 70-1, 152, 153, 154, 155
Egg Island, 19

Ellen Island, 24
Esquimalt, 12
Euclataw. *See* Indians, Lekwiltok
Eulachon, 104, 105
Eulachon oil, trade in, 105, 129-30

Fairbairn Shoals, 44 n. 197
Fanny Bay, 90
Fawcett, Edgar, 93, 95
Fellows, Arthur or Alfred, 95
Fife Point, 56, 58
Findlayson, Roderick, 93
Fishers Channel, 20
Fitzhugh Sound, 19-20, 83
Fort McLoughlin (Bella Bella), 21 n. 67
Fort Rupert, 17; Dawson at, 85-8
Fort Simpson. *See* Port Simpson
Fossils: from Cumshewa Inlet, 41-2; from Fort Rupert, 87; from Houston Stewart Channel, 25; from Moresby Island, 48; from Sandilands Island, 49; from Skidegate Inlet, 48
Fraser Island, 60 n. 277
Fraser Reach, 79
Frederick Island, 163

Garcin Rocks, 24 n. 85
Glaciation, in Georgia Strait, 15
Gnarled Islands, 73
Gold: Chimseyan Lode, Skeena River, 74, 77-8; in Kuper Inlet, 52
Gold Harbour. *See* Kuper Inlet
Goletas Channel, 18, 84, 85
Goudy, Captain, 94
Graham Island, 4
Graham Reach, 79
Grand Prairie, 94
Grappler, 75, 78, 81
Gray, John Hamilton, 92, 94
Gray, Robert, 152
Grenfell Island, 78

Grenville Channel, 78
Grief Island, 80

Haddington Island, 88
Haida, 10, 87
    argillite, 141
    armour, 100
    artifacts collected: from Skidegate, 51; from Tanu and Koona, 41; from Anthony Island, 25; from Bella Bella, 21
    burial, 49-50, 117, 126-7
    canoes, 67, 68, 138-9
    chiefs, 112-14
    coppers, 128-9, 141
    crests, 59, 127-8
    Cumshewa, Chief, 41
    dancing, 45-7, 121-3, 130-2
    dress, 99-101, 130
    face-painting, 101
    farming, 54, 107, 108
    first Whites, 62, 70-1, 146-53
    fishing, 103-6
    food, 67, 103-9
    frontlets, 100-1
    fur trade, 146-53
    gambling, 122-3
    household goods, 133-6
    houses, 46, 109-10, 111-12, 139-40
    labrets, 102
    land title, 166
    language, 3-4, 56, 71, 94, 98
    marriage, 123
    masks, 131
    myths and traditions, 142-6
    names, 125
    on Nass, 77
    ornament, 101-3, 130
    physical description, 98-9
    population, 53, 68, 163, 165
    population decline, 110, 158, 163, 165
    position of women, 123-5

potlatch, 47, 54, 56, 59, 70, 112, 113, 114, 119-21
predecessors, 3
property rights, 111
prostitution, 124
puberty ceremony, 77
rattles, 131-2
religion, 114-19
shaman, 49-50, 60, 115, 116-18, 126-7
Skatz-sai, Chief, 52
Skedan, Chief, 39
slavery, 125-6, 128; social organization, 109-14
soulcatcher, 132-3
tatooing, 97, 102, 128
tools, 136
totem poles, 40, 41, 46, 47, 53, 110, 140-1, 156
Tshimsian influence upon, 121, 144
villages, 109-11, 112, 154-63, 165
warfare, 97
Weah, Chief, 70
weapons, 136-8
Haida (village of). *See* Masset
Hall, Rev. Alfred James, 87
Hall, Robert H., 77
Hamilton, Gavin, 17, 22
Hanington, Dr. E.B.C., 92
Hardy Bay, 87
Harriet Harbour, 26, 29, 30
Harrington, Anna, 11, 23; letter to, 11-12
Harrington, John Eric, 11
Hazelhurst, 94
Hebrew Mine, 81
Hecate Strait, 4
Heillen River, 56 n. 249
Heisterman, Charles H.F., 93
Helmcken Island, 16
Henslung, 151 n. 15
Hibben, T.N., & Company, 95

Hiellen, 70 n. 341
Hillen River, 157
Hornby Point, 24 n. 86
Houghton, Colonel Charles Frederick, 20
Houston Stewart Channel, 24-5
Howe Copper Mining Company, 93
Hudson's Bay Company: and Kuper Inlet gold, 52; in Bella Bella, 80; in Fort Rupert, 86, 87; in Masset, 155; in Port Simpson, 73, 74, 75, 76, 77, 93; and trade in blankets, 52; in Victoria, 10, 92, 93, 95
Hunt, Robert, 85-6, 87
Hunter, Joseph, 94
Huson, Alden Wesley, 17, 88, 94
Huston Inlet, 27 n. 101
Huxley, T.H., 94

Ian Lake, 62 n. 282
Imray, J.F., 155
Indians. *See also* Haida
artifacts, in Alert Bay, 88-9
Bella Bella, 21, 80, 82-3
Boston, Chief, 21
Cheap, Chief, 18, 83, 84
Chilkat Tlingit, 74
Dene, 71, 77, 102, 140
eulachon fishing, 17, 77
farming, around Georgia Strait, 438
in Fort Rupert, 82, 86-7
houses, in Alert Bay, 88
Kaigani, 68, 77
Kitkatla, 82: at Cumshewa, 42
Kokyet (Heiltsuk), 80 n. 394
Kwaiakah, 86
Kwakiutl: potlatch, 9; shaman, 89; tribes, 86
Kwawkewlth, 86
Lekwiltok, 15, 16, 83, 86
Makah, 104
Mamalillikulla, 86

Matilipi, 86
memorial to Chief Boston, 21
mourning ceremony, Nawitti, 84
myths, in Bella Bella, 82-3
Nakwakto, 86
Nawitti, 83, 84, 86
Nimkish, 17, 86
Nootka, 105
Nsawataineuk, 86
population decline of, 83
potlatch: Alert Bay, 89; Nimkish, 17
sea otter hunt, 54
Tenakteuk, 86
Tlawitsis (Turnour Island), 86
Tongas, 77
totem poles, in Nawitti, 84
trade in eulachon oil, 42-3
Tsimshians, 77, 87: influence upon Haida, 121, 144; at Masset church, 59; origin of, 77; potlatch, 120-1; religion, 118-19; sea otter hunt, 54; slavery, 128
Wewayakum, 86
Iphigenia, 152
Island Bay, 458, 459

Jalun River, 70 n. 343
Jewell Island, 49 n. 211
Juan Perez Sound, 32 n. 120
Juskatla Inlet, 60 n. 276
Juskatla Narrows, 60 n. 276

Kaigani, 4, 155, 164
Kaisan, 52 n. 232, 53, 160
Kate, 85
Kayang, 63 n. 288, 66, 155
Keith, J.C., 12
Kennedy Island, 76, 78
Kerouard Islands, 24 n. 83
Kingui Island, 43 n. 194
Kitkatla, 130
Kiusta, 72 n. 351, 154

Klas-kwan Point, 155
Klashwun Mountain, 70 n. 336
Klashwun Point, 70 n. 342
Klemtoo Passage, 80
Klue, 32, 39, 41
Klunkwoi Bay, 35 n. 150
Knights Inlet, 17
Kogals-Kun, 62 n. 284
Kokyet, 80 n. 394
Koona, 39, 40, 41, 43, 157, 161
Koskimo, 86, 87, 91
Kung, 66, 70 n. 334, 152, 155
Kunga Island, 37 n. 158
Kuper Inlet, 4, 25, 43, 48, 51, 52, 93, 104, 106
Kwaiakah, 86
Kwude, Dr., 60 n. 271
Kynumpt Harbour, 20, 23

Lama Passage, 20, 23, 83
Landale, John J., 91
Langara Island, 4, 62, 69, 70-2, 74, 153, 154
LaPérouse, Compte de, 147
Laskeek Bay, 36 n. 147
Lauder Point, 72 n. 353
Laurel Point, 92
Lawn Hill, 55, 57
Leighton, Mr., 81
Lepas Bay, 72 n. 352
Lewis, Herbert George, 10
Lina Island, 49
Logan Inlet, 37 n. 163
Long Inlet, 50 n. 217
Lucy Island, 71, 73, 151 n. 14, 154
Lyell Island, 37 n. 155

Maast Island, 63, 156
McClinton Bay, 61 n. 280
McCoy Cove, 39
Macdonald, Senator William John, 11, 12, 93, 95
McGregor, Andy, 4, 45, 48

McKay, Hugh, 39, 43
McKay, Joseph William, 20, 22, 74, 75, 79, 81, 93
McKay Reach, 79
McKays Cove. *See* McCoy Cove
McKenzie, Alexander, 152
McLellan Island, 44 n. 198
McLoughlin Bay, 20
McPherson Point, 71 n. 348
Madden William F., 77
Makah, 104
Malcolm Island, 18, 88
Mamalillikulla, 86
Mamin River, 60, 64 n. 298, 65
Mamin village, 60 n. 278
Maquina (Maquilla), Chief, 103
Marchand, 140
Martineau, Miss, 95
Masset, 57, 63 n. 287, 149 n. 11, 155; Dawson at, 57, 58-60; from Skidegate, 55-8; to Cape Knox, 66-73
Masset Inlet, 61 n. 279, 64, 65, 155-6
Masset Sound and Inlet, Dawson at, 60-6
Mathers Creek, 44 n. 199
Maude Island, 48, 49, 51, 160
Maurelle, 153
Meares, 147, 149, 154
Milbank Sound, 20, 80
Mittlenatch Island, 15
Moffatt, Hamilton, 41
Moore Head, 24 n. 84
Moresby Lake, 52 n. 231
Mount Albert Edward, 542
Mount Griffin, 73 n. 357
Muirhead & Mann, 92
Murchison Island, 34 n. 128
Mustoo, 162

Naden Harbour, 66, 67 n. 322
Naden River, 66
Nakwakto, 86
Nanaimo, 91; Dawson at, 14-5

Nawitti, 84, 86; from Port Simpson, 76-84
New Gold Harbour, 48, 53, 160
Newcastle Ridge, 16 n. 45
Newittie, 109
Nimpkish Lake, 17 n. 54
Nimpkish River, 17
Ninstints, 25, 54, 161
Nodales Channel, 16
North Island. *See* Langara Island
Nsawataineuk, 86

Offut, Martin H., 4, 57, 59, 66, 69
Ogden Channel, 76
Olympic Mountains, 14
Otter, 15, 17, 20, 21, 22, 57, 63, 78, 80, 87

Parry Passage, 73 n. 354, 150 n. 13, 151, 152, 154-5
Pender Island, 89
Pérez, Juan, 4
Pillar Rock, 89
Pitt Island, 78
Plumper Bay, 16
Poole, Francis, 26, 27, 30
Poole Point, 27 n. 103
Pooley, Charles Edward, 12
Port McNeill, 88 n. 422
Port Simpson, 74-6, 109, 118; from Cape Knox, 72-3; to Nawitti, 76-84
Portland Inlet, 73, 76
Portland Island, 14
Powell, Dr. Israel Wood, 12
Prince of Wales Island, 3
Prince of Wales Range, 16

Qauatse River, 87 n. 414
Quadra Island, 86 n. 412
Quadra Rocks, 24
Quatsino Sound, 84, 109
Queen Charlotte, 160, 163
Queen Charlotte Islands, 3; Dawson at, 23-73; from Victoria, 13-23

Index

Race Rock, 95
Redtop Mountain, 35 n. 144
Reef Island, 49-50
Richards, Albert Norton, 12, 94
Richardson Island, 36 n. 146
Richardson, James, 11, 15
Rivers Inlet, 19
Robinson, Captain and Mrs. James, 12
Robson Bight, 17
Robson, John, 94
Rockfish Harbour, 38 n. 164
Roscoe, Francis James, 93
Rose Harbour, 24 n. 89
Rose Inlet, 25 n. 89
Rose Point, 58, 70, 146
Rose Spit, 56, 58

Saanich Inlet, 92
Sabiston, John, 11, 15, 22, 92
Safety Cove, 20, 83
Salmon River, 17
Sandilands Island, 49 n. 214
Scudder Point, 32 n. 116
Sea Otter Island, 19
Seaforth Channel, 81
Section Cove, 32 n. 114
Selwyn, A.R.C., 88
Seymour Narrows, 15, 89 n. 429
Shadwell Passage, 83
Ships: *Dakota*, 9, 95; *Grappler*, 75, 78, 81; *Hazelhurst*, 94; *Iphigenia*, 152; *Kate*, 85; *Otter*, 15, 17, 20, 22, 57, 63, 78, 80, 87; *Queen Charlotte*, 160, 163; *Susan Sturgis*, 70 n. 338; *Vancouver*, 58 n. 258; *Wanderer*, 11, 13, 15, 22; *Washington*, 152
Shushartie Bay, 18
Skaat Harbour, 33 n. 123
Skatz-sai, Chief, 52
Skedan, Chief, 39
Skedan. *See* Koona
Skedans Bay, 40 n. 182
Skedans Island, 43 n. 196

Skeena River, 76-7
Skidegate, 4, 103, 157-60
Skidegate Channel, 48, 50-1, 54-5
Skidegate Inlet, 4, 164; Dawson at, 44-53, 55; from Cape St. James, 24-44, 48; to Masset, 55-7, 57-8
Skidegate Landing, 45 n. 208
Skidegate Mission. *See* Skidegate
Skincuttle Inlet, 26
Skincuttle Islands, 4
Slate Chuck Creek, 55, 64
Smallpox, 4
Smith, J. McB., 4, 45, 48, 51, 53
South Cove, 26 n. 98
Spencer, S.A., 95
Spiller Channel, 81 n. 395
Spit Point, 44, 45
Sproat, Gilbert Malcolm, 90
Stikine River, 77
Sturt, Captain D.E., 95
Suquash coal fields, 87
Susan Sturgis, 70 n. 338
Susk, 163
Sutton, William, 91
Swan, J.G., 140
Swan Bay, 29 n. 105
Sydney Island, 14

Tangle Cove, 30, 31
Tanu, 32, 161
Tar Rock, 37 n. 156
Tasu Sound, 43, 162
Tenakteuk, 86
Tennyson, Alfred Lord, 85
Texada Island, 95
Tlawitsis (Turnour Island), 86
Tlell, 157
Tobacco (Haida), 108
Tolmie, Dr. William Fraser, 92, 94, 95, 97, 164
Tolmie Channel, 78
Torrence, J. Fraser, 9
Torrens Island, 49 n. 211, 53 n. 236
Tow Hill, 56, 67, 72, 145

Towustasin Hill, 67 n. 324, 145
Trial Island, 14
Trounce Inlet, 51 n. 225, 55 n. 244
Tsmishian. *See* Indians, Tsimshian

Ucultas. *See* Indians, Lekwiltok
Ut-te-was. *See* Masset

Vancouver, 58 n. 258
Vancouver, George, 152
Vere Cove, 89 n. 427
Victoria, Dawson at, 9-14, 92-5
Virago Sound, 66, 70, 72, 73, 155

'W.B.,' 147
Walbran Island, 19 n. 61
Walkem Island, 89 n. 428
Wallace Island, 92 n. 442
Wanderer, 11, 13, 15, 22
Ward, William C., 95
Washington, 152
Watt, M., 90
Watun River, 60 n. 265
Weah, Chief, 70 n. 337, 156
West Narrows (Skidegate Channel), 50 n. 222, 54
Weston, Thomas Chesmer, 11
Wewayakum, 86

White, James, 11
Whitecliffe Island, 76
Williams, John, 15, 23, 24, 27, 28, 50, 92
Wimbledon Mountain, 78
Woodcock, William H., 52, 93
Woodcock's Landing. *See* Inverness
Woodcocks Fishery. *See* Skidegate Landing
Work Channel, 77
Work Island, 80
Work, John, 156, 157, 158, 162, 164, 165
Wrangel, 164
Wright, Rev. Henry Press, 95
Wright Sound, 78, 79

Yakoun, 156
Yakoun Lake, 64 n. 294
Yakoun River, 60 n. 270, 64 n. 293
Yan, 63 n. 289, 66, 155
Yatze, 66, 67 n. 503, 73, 152, 155
Yolk Point, 78
Yugan, 57
Yukan Point, 57 n. 252

Zayas Island, 73

Typeset in Palatino by Arifin Graham, Alaris Design

Printed and bound in Canada by D.W. Friesen & Sons Ltd.

Copy-editor: Stacy Belden

Proofreader: Joanne Richardson